Planetary Mine

Territories of Extraction under Late Capitalism

Martín Arboleda

V
VERSO
London • New York

NAKHODKA
BAYUQUAN
TANGSHAN (JINGTANG)
TIANJIN XIN GANG
QINGDAO GANG
RIZHAO
LIANYUNGANG
ZHOUSHAN

KHAWR
FAKKAN
AL RAYYAN
TERMINAL
SITRAH

ENNUR

FANG-CHENG

APRA
HARBOR

PORT HEDLAND
PORT WALCOTT

--------- Destination of *Capesize* bulk carriers departing from Chilean Ports in 2015
Coastal boundary

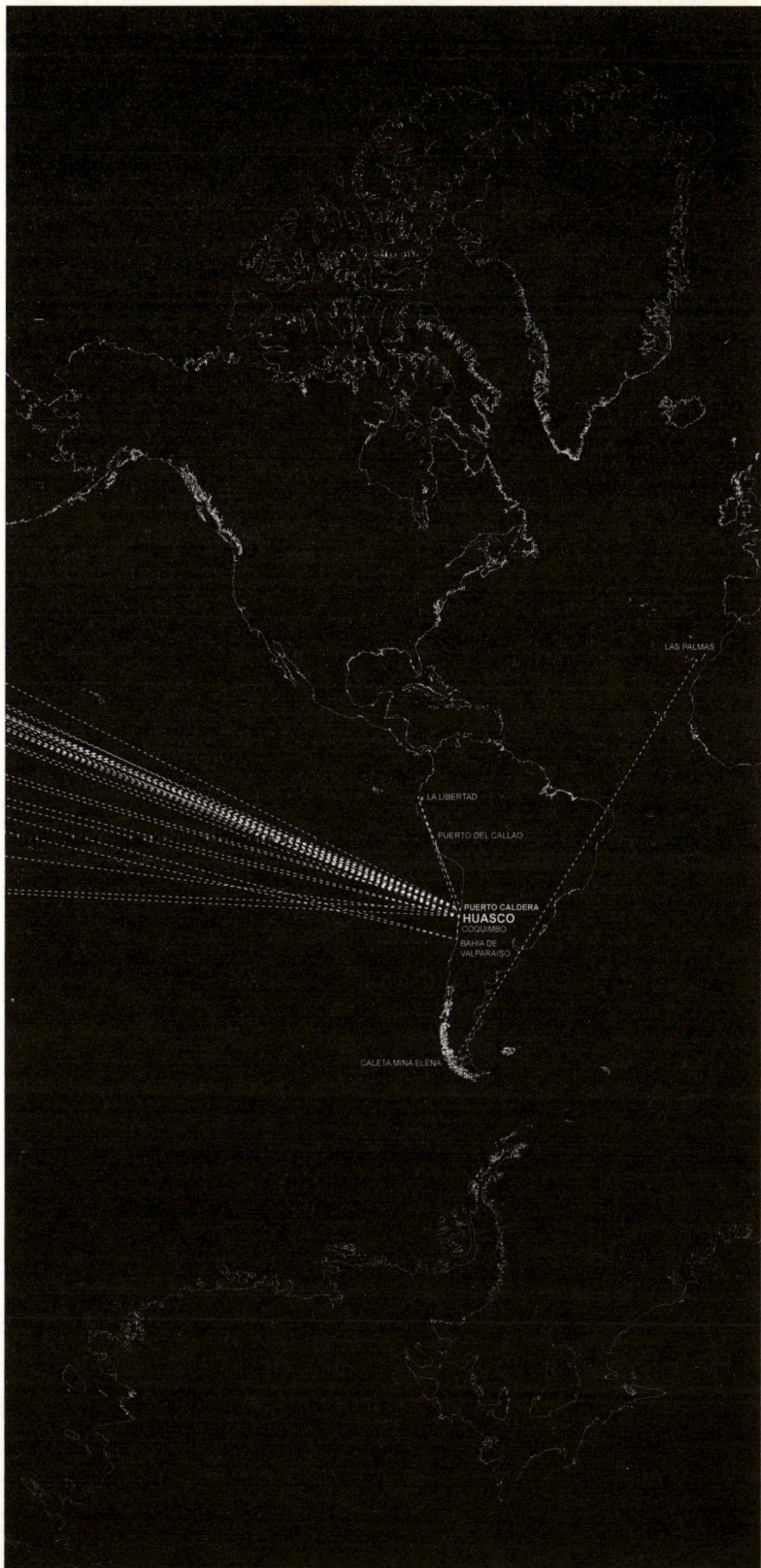

LAS PALMAS

LA LIBERTAD

PUERTO DEL CALLAO

PUERTO CALDERA
HUASCO
COQUIMBO

BAHIA DE
VALPARAISO

CALETA MINA ELENA

First published by Verso 2020
© Martín Arboleda 2020
The map on pages vi–vii was produced by Carla Ferrer-Llorca with
data from the University College London's Energy Institute.

The moral rights of the author have been asserted

1 3 5 7 9 10 8 6 4 2

Verso
UK: 6 Meard Street, London W1F 0EG
US: 20 Jay Street, Suite 1010, Brooklyn, NY 11201
versobooks.com

Verso is the imprint of New Left Books

ISBN-13: 978-1-78873-296-3
ISBN-13: 978-1-78873-295-6 (LIBRARY)
ISBN-13: 978-1-78873-297-0 (UK EBK)
ISBN-13: 978-1-78873-298-7 (US EBK)

British Library Cataloguing in Publication Data
A catalogue record for this book is available from the British Library

Library of Congress Cataloging-in-Publication Data
A catalog record for this book is available from the Library of Congress

Typeset in Minion Pro by Hewer Text UK Ltd, Edinburgh
Printed and bound by CPI Group (UK) Ltd, Croydon CR0 4YY

Europe is no longer the center of gravity of the world. This is the significant event, the fundamental experience, of our era.

Achille Mbembe, *Critique of Black Reason*

The machine itself makes no demands and holds out no promises: it is the human spirit that makes demands and keeps promises. In order to reconquer the machine and subdue it to human purposes, one must first understand it and assimilate it.

Lewis Mumford, *Technics and Civilization*

I see the mountains fall, open up the territory in angry grayish cavities, the desert, the transitory houses. The mineral on fire and shaped and handled became military ingots, battalions of merchandise. The ships departed. Wherever copper arrives, utensil or wire, no one touching it will see the rugged solitudes of Chile, or the small houses on the edge of the desert . . .

Pablo Neruda, *Ode to Copper*

Table of Contents

Table of Figures   ix

Acknowledgments   xi

1. OPENINGS   1
The Mine as Transnational Infrastructure

2. EMPIRE   35
Resource Imperialism after the West

3. LABOR   75
Bodies of Extraction and the Making of Urban Environments

4. CIRCULATION   109
State Power and the Logistics Turn in the Extractive Industries

5. EXPERTISE   140
Technocracy and Expropriation

6. MONEY   175
Debts of Extraction

7. STRUGGLE   206
Plebeian Consciousness and the Universal Ayllu

8. EPILOGUE   243
Toward an Emancipatory Science in the City of Extraction

Index   261

Table of Figures

Figure 1. World exports of mining machinery (left axis),
and China's imports of minerals (right axis).   3

Figure 2. Total world exports of machinery and
electrical equipment during the 1988–2015 period.   57

Figure 3. China's imports of minerals and stone/glass
during the 1995–2015 period, in billions of US dollars.   62

Figure 4. China's imports of foodstuffs during
the 1995–2015 period, in billions of US dollars.   64

Figure 5. Distribution of services provided
to the mining industry in Chile.   90

Figure 6. Employment growth in Santiago according to
occupational group, 1992–2002, expressed in thousands.   93

Figure 7. Share of subcontracted workers in
Chile's mining industry 1975–2004.   95

Figure 8. Credit allocations in Vallenar and Barrick
Gold's capital expenditures on the Pascua Lama project.   198

Figure 9. Financial intermediation in Vallenar
and sales of hotels and restaurants in Vallenar.   199

Acknowledgments

This book interrogates the political life of infrastructures of extraction, especially at a time when new generations of machines and integrated industrial technologies actively reweave the textures and environments of everyday life, and the Atlantic Ocean ceases to be the pivot upon which the modern world has always turned. If "to be one is to *become with* many," as Donna Haraway suggests, this book is then far from being a conclusive statement; it is rather the embodied manifestation of an ongoing dialogue with many individuals who—like myself—have abhorred and felt frustrated, but also captivated and heartened, by this period of great historical transformations. I have had the good fortune of being surrounded by the best colleagues, critics, and friends that anyone could have. I am deeply grateful to Neil Brenner for his encouragement, support, and guidance throughout the whole process of writing this book. During my time at Harvard University, Neil read several revisions of chapter drafts despite his incredibly busy schedule; he also mobilized tirelessly within the university and beyond so that I could have the time, space, and infrastructure required for writing this book. This book has also benefited greatly from the talent and intelligence of my colleagues at the Urban Theory Lab. Mariano Gómez-Luque, Julia Smachylo, Daniel Ibáñez, Ghazal Jafari, Nikos Katsikis, Mike Chieffalo, Ayan Meer, and Tamer Elshayal, have been excellent colleagues and interlocutors, and I have learned so much from them. Other colleagues at Harvard to whom I am very grateful are Stuart Schrader, Rachel Meyer, Amanda Miller, and Carla Ferrer-Llorca.

This book also owes a huge debt of gratitude to several of my colleagues at the University of Manchester. Despite having completed his role as my PhD supervisor years ago, Greig Charnock has continued to read innumerable drafts and chapter revisions, always getting back to me quickly and with incisive and thoughtful feedback. Also, Greig's efforts to revisit the New International Division of Labor (NIDL) thesis have been revelatory, and figure centrally in this book. Adrienne Roberts and Carl Death also provided very relevant guidance and mentoring. My former colleagues at the Society and Environment Research Group in Manchester, who went on to become close friends, have also contributed in one way or another to the ideas contained in this book. They are Julie Ann De Los Reyes, Melissa García-Lamarca, Creighton Connolly, and Daniel Banoub. Japhy Wilson and Tom Purcell have also been close colleagues and interlocutors during the time I have spent writing this book, having also offered very relevant feedback on chapter drafts. Ilias Alami made very helpful and detailed comments on one of the most difficult chapters, and for this I am very grateful.

Other fellow travelers and colleagues beyond the universities that I have been directly affiliated with, have also contributed in different ways to this intellectual journey. Alex Loftus has taken the time to read and discuss work in progress of my research across several of its iterations. His insistence on the relevance of everyday experience and situated knowledge shaped in crucial ways my own ideas about extraction and global capitalism since early on in my doctoral research. Gastón Caligaris was generous enough to send me some of the books that have emerged from the innovative theoretical work produced by the Buenos Aires–based Centro para la Investigación como Crítica Práctica (CICP). Jason W. Moore has been very supportive throughout the whole process, and invited me to present some of the ideas of this book at the 2017 World-Ecology Conference in Binghamton, United States. Nicole Aschoff helped me to think through the intersections between the mining industry and the electronics industry. Mazen Labban introduced me to Enrique Dussel's pathbreaking study on Marx's manuscripts of 1861–63, which was formative as I revised some of the key chapters of the book. Mazen's own work has also been a source of inspiration, and this book is a collegial attempt to take forward in new directions his lucid call to rethink mineral extraction beyond the extractive industries.

My field research in Chile also benefited greatly from the generosity of many individuals. I am particularly thankful to Patricia Álvarez, Valeska Urqueta, Cristian Ochoa, Andrea Cisternas, Juan Carlos Labrín, Soldad Fuentealba, Jorge Guerrero, Javier Karmy, Consuelo Infante, and Lucio Cuenca. Mark Muller, Ixent Galpin, Andy Higginbottom, and Richard Solly, from the London Mining Network, were also very friendly and keen to help with the development of this research, when it was still at an early stage. The Urban Studies Foundation (USF) generously funded this research with a postdoctoral fellowship that allowed me to have almost exclusive dedication to the book project for a period of two years. More than a mere "funding body," the Urban Studies Foundation is an institution enlivened by wonderful individuals who are genuinely committed to fostering the professional and intellectual development of their fellows. I especially want to thank Chris Philo, John Flint, Ruth Harkin, Donald McNeill, Neil Gray, and Joe Shaw, for being so generous, open, and flexible during the time I was a USF fellow. At Verso, I thank Sebastian Budgen for his interest in this book, and Duncan Ranslem for all the support he provided during crucial moments of copyediting and production.

Moreover, I also thank several colleagues in Santiago de Chile for their encouragement, feedback, and for the stimulating conversations during the years that I spent writing this book. They are Michael Lukas, Carlos de Mattos, Luis Andueza, Ernesto López-Morales, Tomás Ariztía, Rodrigo Cordero, and Sebastián Ureta. I also thank all my colleagues at the School of Sociology of Universidad Diego Portales, for warmly welcoming me to their intellectual community. Some sections of chapter 6 have previously appeared published as an article titled "Financialization, Totality and Planetary Urbanization in the Chilean Andes," in *Geoforum* (2015), so I thank Elsevier for granting permission to reproduce them here. Finally, I thank my extended family, scattered across Colombia, Chile, and Panama, for years of endless support and affection. And crucially, this book simply would have never existed without Mela, my wife, comrade, and fellow traveler. Not only did she have the patience to discuss with me version after version of book chapters, but she also brought joy and loving support to this arduous intellectual journey. Her strength and graciousness have been a powerful source of inspiration for everything I do. I dedicate this book to her and to our sons Lucas and Tomás.

The idea to write this book originated during my final months as a PhD student at Manchester, and especially after the birth of my son Lucas under the United Kingdom's system of universal public health service provision. Due to some complications at birth, little Lucas was taken to a neonatal intensive care unit because he was unable to breathe. He was immediately connected to a ventilator and to a whole range of tubes, cables, and screens that constantly monitored his bodily rhythms and the various medicines that were being pumped into his organism, while a group of nurses and doctors cared for him affectionately and professionally. Seeing this ensemble of sophisticated machines and highly skilled health professionals working day and night to save the life of Lucas—the son of two immigrants who had no money in their bank accounts—brought closer to home the central argument of this book: that science and technology will realize the plenitude of their soothing and creative powers once they break the spell that private property has cast upon them. That if this happens, humans and machines might overcome their existence as personifications of the phantom-world of capitalist economic abstraction (whether as laborers, debtors, fixed capital, and so forth) and thrive in the affirmation of *their own* sensuous, textured, and concrete particularities. A world of liberated labor and liberated technology is worth fighting for, and popular struggle against the supply chain of extraction is already leading the way in that direction. I hope that this book honors such struggles, and that it provides a sound conceptual and empirical basis for understanding our present technological reality in order to transform it.

1. OPENINGS

The Mine as Transnational Infrastructure

Introduction

In the early hours of July 21, 2015, nearly 300 subcontracted workers demanding better conditions decided to occupy El Salvador, a copper mine located in the Atacama region of northern Chile. After partially blocking the road to Diego de Almagro—the neighboring town—and shutting down activities at the extraction site, the picketing workers used three Scoop loader machines (heavy haulage equipment employed to move large volumes of rocks) to erect barricades in order to protect themselves from the police crackdown that they viewed as impending. They were not mistaken. On the evening of July 23, a heavily armed contingent of around 120 police and special forces in full antiriot gear arrived and a raging battle broke out. Stones thrown from the barricades set up by these temporary and precarious workers clashed with the rubber bullets and tear gas canisters of the police. After several hours of confrontation, the striking workers mobilized the Scoop loaders they had used to erect the barricades to push back the police's advance and also to shield themselves from escalating police violence. They had expected considerable repression, but the furious backlash unleashed against them was simply beyond anyone's imagination. It was as if the police had been possessed by a strange and overwhelming power. "They were pointing infrared beams at us, we were suffocating in tear gas; airgun pellets and rubber bullets were being fired indiscriminately and heavily injuring our comrades," recalls a worker. After the Scoop machines were drawn into the standoff, the police began to fire live ammunition against the blockade. One of the shots took the life of Nelson Quichillao, who bled to death amid the bitter tears and astonishment of his coworkers, who could barely believe what had just happened.[1]

For more than fifteen years, since he was eighteen years old, Quichillao had been working for the mining industry, but he had never managed to obtain a stable work contract that offered him health insurance, paid vacations, and access to a retirement plan, among other benefits that come with salaried work under Chile's labor regulations.[2] Though it was hardly covered by the mainstream media, Nelson Quichillao's death is far from being an insular or fragmentary event in the swirling complexities of the historical process. In fact, it holds the secret to one of the most fundamental world-historical shifts in late modern society: the staggering acceleration of automation and the concomitant replacement of human laborers with intelligent machines, enabled by a recent leap forward in the whole productive technology of capitalism.[3] The tendency to increase the organic composition of capital—that is, the ratio of automated labor to living labor—has been an intrinsic feature of the capitalist mode of production since machinofacturing (the production of machines by machines) became the underlying technical foundation of large-scale industry. Since the turn of the twenty-first century, however, we appear to be witnessing a new stage in the historical struggle of capital against labor, brought about by a new generalized architecture of social production. As figure 1 illustrates, the tendency of the mining industry to become more capital-intensive experienced a dramatic turning point when world exports of mining and construction machinery

---

1 Clinic Online, "Memoria sin justicia: el monumento a Nelson Quichillao," 2016, theclinic.cl.

2 *El Mostrador*, "Nelson Quichillao: el fatal destino de un minero subcontratado," July 31, 2015, elmostrador.cl.

3 The murder of striking miners by police forces is a growing trend across Latin America and beyond, especially as working conditions deteriorate abruptly and rapidly. In one of the most violent of these standoffs, forty-four mine workers striking for improved pay were massacred by the South African police in 2012 while protesting at the Marikana mine; seventy-eight others were seriously injured by gunshot wounds. Stephen Graham, *Vertical: The City from Satellites to Bunkers* (New York: Verso, 2016), 60.

increased from $17 billion in 2002, to $65 billion in 2012. In Chile, the life and death of Nelson Quichillao thus came to symbolize the plight of temporary and subcontracted workforces, whose ranks have swollen as the mining industry becomes ever more "smart," "flexible," and "autonomous."[4] In April of 2016, a trade union named Frente de Trabajadores Nelson Quichillao was created in order to defend subcontracted workers against layoffs and further labor casualization by the mining industry.

Figure 1 World exports of mining machinery and China's imports of minerals

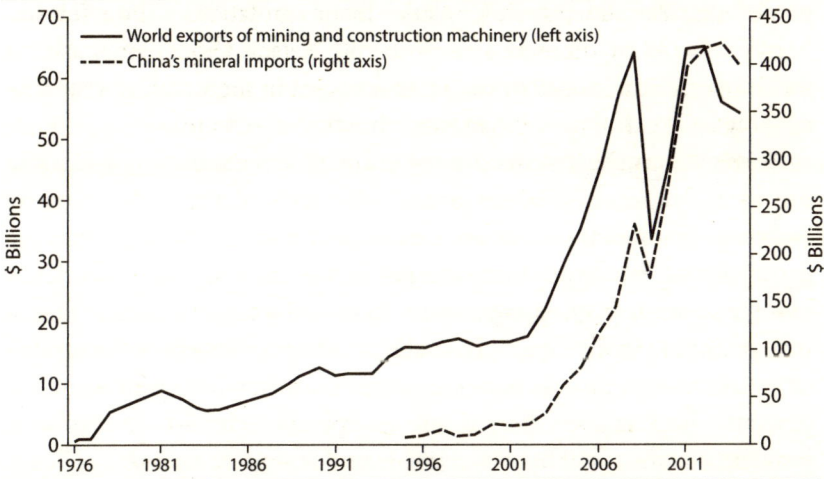

Source: World Integrated Trade Solution and Atlas of Economic Complexity.

Although in the popular imagination mining is usually considered to be a rudimentary activity, the degree of technological sophistication that mediates the extraction of minerals from the subsoil in the twenty-first century is nothing short of astonishing. Innovations in artificial intelligence, big data, and robotics have allowed mining companies to introduce automatic trucks, drills, shovels, and locomotives to the stages of the production process. Some of these sophisticated machineries—most notably trucks and shovels—are not remotely controlled; they are fully robotized, which means that they can operate twenty-four hours a

---

4 *Izquierda Diario*, "El Frente de Trabajadores Nelson Quichillao y la necesaria unidad de acción para enfrentar los despidos en la minería," 2016, laizquierdadiario. com.

day, seven days a week, without direct human intervention.[5] The introduction of geospatial information systems to mineral forecasting and geological surveying has also allowed engineers to produce highly accurate representations of the subsurface, making the extraction of low-grade ore bodies profitable for the first time in history. Mines that had been abandoned are therefore being reopened, and deposits that were deemed uneconomical are being transformed into large open-cast extraction sites across all stretches of the planet.[6] Increasing spatial separation of extraction and manufacturing has also pushed the mining industry toward greater functional integration with the port and shipping industries. An erstwhile insular focus on the extraction site has been gradually superseded by a deliberate organizational emphasis on the supply chain, understood as an integrated logistical system that encompasses extraction, processing, smelting, and transport.

This book argues that the mining industry's recent technological and organizational modernization transcends mere shifts in the intensity and scale of mineral extraction. The *planetary mine*, I will argue, is the geography of extraction that emerges as the most genuine product of two distinct yet overlapping world-historical transformations: first, a new geography of late industrialization that is no longer circumscribed to the traditional heartland of capitalism (i.e., the West), and second, a quantum leap in the robotization and computerization of the labor process brought about by what I will term the *fourth machine age*. Since the long sixteenth century, resource booms have been understood as the direct outcome of relations of unequal exchange between an imperial power and a dependent periphery, in the context of a Eurocentric capitalist system. However, the technological modernization and industrial upgrading taking place across the global South after the 1980s—especially in East Asian economies—have decentered the geography of

---

5   COCHILCO (Comisión Chilena del Cobre), *Biolixiviación: desarrollo actual y sus expectativas* (Santiago, Chile: 2009); Corporación Nacional del Cobre (Codelco), *Codelco Digital*. Power Point presentation, March 3, 2011, codelco.com; Rio Tinto, *Next-Generation Mining: People and Technology Working Together* (Melbourne: Rio Tinto, 2014); Tom Simonite, "Mining 24 Hours a Day with Robots," *MIT Technology Review*, December 29, 2016, technologyreview.com; Fundación Chile, *Desde el cobre a la innovación: Roadmap tecnológico 2015–2035* (Santiago, 2016).

6   Geoff Manaugh, "Infinite Exchange," in David Maisel, *Black Maps: American Landscape and the Apocalyptic Sublime* (Göttingen, Germany: Steidl, 2013); Richard Swift, 'Stop the Gold Rush!" *New Internationalist* 475 (September 2014).

large-scale industry, destabilizing traditional metageographical categories of core/periphery and even of global North/global South.

Throughout this book, I show how geographies of extraction have
become entangled in a global apparatus of production and exchange
that supersedes the premises and internal dynamics of a proverbial
world system of cores and peripheries defined exclusively by national
borders. The main contention of this book, therefore, is that the mine is
not a discrete sociotechnical object but a dense network of territorial
infrastructures and spatial technologies vastly dispersed across space. I
build upon Mazen Labban's notion of the *planetary mine* as one that
vastly transcends the territoriality of extraction and wholly blends into
the circulatory system of capital, which now transverses the entire geography of the earth.[7] The technological basis of contemporary mining,
Labban notes, has blurred the boundaries between manufacturing and
extraction, waste and resources, biologically and nonbiologically based
industries. This, in his view, warrants the reconsideration of extractive
industries beyond the mere wresting of minerals from the soil. I therefore build upon recent approaches that have considered spaces of extraction to also include logistical infrastructures, transoceanic corridors,
networks of financial intermediation, and geographies of labor. The
reorganization of the mining industry into global supply chains engen-
ders novel modalities of state power and capitalist imperialism and
yields a new territoriality of extraction whose immanent content cannot
be fully elucidated by the *loci classici* of state-centric concepts of political
economy, such as resource curse, dependency, imperialism, and so
forth.

The methodological nationalism that informs most studies of extraction is analytically debilitating because it obfuscates how deeply inter-
twined global supply chains and sprawling urban systems are in the
sociometabolic production and reproduction of the mine; it is also *politically counterproductive* because it pits the workers and communities of
"resource-rich" countries against those of the manufacturing centers,
when in fact they all share increasingly common conditions of existence. On this basis, this book makes three central and interrelated

---

7 Mazen Labban, "Deterritorializing Extraction: Bioaccumulation and the
Planetary Mine," *Annals of the Association of American Geographers* 104, no. 3 (2014):
560–76.

arguments. First, it insists on the fact that the concrete determinations that produce spaces of extraction are not relations of unequal exchange and dependency, but *the production of relative surplus value at the world scale and the reproduction of the international working class as a fragmented, polarizing, yet unitary whole or industrial organism.* In concert with recent work on the new international division of labor and on the new geographies of advanced industrialization, it uses resource extraction as an analytical entry point to theorize uneven geographical development after the Western phase of capitalism. Second, it argues that the determination of capital as a genuinely global form of social mediation does not entail falling into the hyperglobalist fallacy that posits the erosion and withering away of state sovereignty. The political authority that underpins the international movement of capital continues to be mediated nationally; hence the existence of the planetary mine signals the emergence of a more coercive, centralized, and authoritarian configuration of late neoliberal statecraft.

Finally, *Planetary Mine* contends that understanding the manifold determinations that produce landscapes of extraction in historically and geographically specific ways necessitates a dialectical theory of praxis, one that sets out from the material conditions in which life itself is produced and reproduced. As the young Marx argued, "Sense perception must be the basis of all science."[8] It is by interrogating sensuous practice that we become better positioned to grasp the manifestations of the mystifying forms, the immanent rhythms, and the inner contradictions of the capitalist economy and bourgeois society in its totality. At the heart of this book's intellectual project is therefore an attempt to reclaim "form-analysis Marxism," a strand of critical thought that has remained peripheral to the development of historical-geographical materialism and urban political ecology but has much to offer to these vibrant fields of inquiry.[9] A focus on the *forms* or *modes of existence of*

---

8   Karl Marx, *Economic and Philosophic Manuscripts 1844* (New York: Dover, 2007 [1844]), 111.

9   For formative statements, see Simon Clarke (ed.), *The State Debate* (London: Macmillan, 1991); Werner Bonefeld, Richard Gunn, and Kosmas Psychopedis (eds.), *Open Marxism: Dialectics and History*, Vol. 1 (London: Pluto Press, 1992); Hans-Georg Backhaus, "Between Philosophy and Science: Marxian Social Economy as Critical Theory," in Bonefeld et al. (eds.), *Open Marxism* Vol. 1; Moishe Postone, *Time, Labor, and Social Domination: A Reinterpretation of Marx's Critical Theory* (Cambridge: Cambridge University Press, 2003 [1993]); Werner Bonefeld, *Critical Theory and the*

capital rather than on "structures" allows us to supersede subject-object dualism, and henceforth capture the universal content that is expressed through the unfolding of concrete practices and things.[10] To the extent that it seeks to decipher global processes through their concrete manifestation in the situated, affective fabrics of human and nonhuman existence, this mode of theory-building strongly resonates with the aspirations of approaches such as standpoint feminism and minor theory.[11]

Under a categorical approach that roots the dialectical critique in experience, the gendered and racialized migrant toiling in the popular economies of a mining town no longer appears as an isolated fact of social reality, but embodies deep reconfigurations in the social composition of the global working class and the relative surplus population; the robotized systems of extraction wresting copper from the bowels of the Andes begin to reveal the metabolic process that underpins the startling growth of megafactories and industrial cities in East Asia; the credit cards in the hands of peasants living near remote extraction sites crystallize fragments of the hulking figure of "casino capitalism" and its complex global architectures for the organization of monetary flows. But perhaps most importantly, an emphasis on forms also foregrounds

---

Critique of Political Economy: On Subversion and Negative Reason (London: Bloomsbury, 2016); Riccardo Bellofiore and Tommaso Redolfi Riva, "The *Neue Marx-Lektüre*: Putting the Critique of Political Economy Back into the Critique of Society," *Radical Philosophy* 189 (2015): 24–36; related approaches in Latin America include Enrique Dussel, *Towards an Unknown Marx: A Commentary on the Manuscripts of 1861–63* (New York: Routledge, 2001 [1988]); Juan Iñigo Carrera, *El Capital: Razón Histórica, Sujeto Revolucionario y Conciencia* (Buenos Aires: Imago Mundi, 2013 [2003]); Guido Starosta, *Marx's Capital, Method and Revolutionary Subjectivity* (Leiden: Brill, 2015).

10   *Form-analysis Marxism* broadly refers to a subterranean rubric of dialectical critique that encompasses the work of authors in the traditions of Open Marxism, the *Neue Marx-Lektüre*, and related approaches in Latin America, notably those of Enrique Dussel and of the scholars associated with the Centro para la Investigación como Crítica Práctica, based in Buenos Aires, Argentina. Broadly understood, these traditions frame the Marxian critique of political economy as an interrogation of the alienated forms of social mediation that are historically specific to modern, capitalist society. Despite some internal divergences, these currents have in common the fact that they reject the methodological separation of politics and economics typical of structural variants of Marxism and emphasize Marx's treatment of alienated labor, fetishization, and alien objectivity in his mature work.

11   Sandra Harding (ed.), *The Feminist Standpoint Theory Reader: Intellectual and Political Controversies* (New York: Routledge, 2004); Cindi Katz, "Towards Minor Theory," *Environment and Planning D: Society and Space* 14 (1996): 487–99.

the impermanence of things and the contingent nature of reality. Thus, in the alienated movement of the clockwork mechanical systems of extraction—themselves a form of existence of capital as it transitions through its phases—we can also begin to perceive the early stirrings of a future society where technology no longer presents itself as a hostile, quasi-autonomous power, but can instead irradiate and nurture life. An approach that takes seriously the analysis of *modes of existence*, Postone argues, necessarily points toward such alternative futures, where content is stripped of its distorting capitalist forms and can finally come into its own.[12]

Natural Resource Frontiers after the Western Phase of Capitalism

Europe, Achille Mbembe suggests, has ceased to be the structuring center of human civilization, and this fact constitutes the most fundamental experience of our time.[13] The downgrading of Europe into one among several other world provinces, Mbembe considers, opens up new possibilities and horizons for critical thought. Perhaps one of the most relevant of these horizons consists in identifying the attributes of capital as it transforms into a sociomaterial form of life whose scale is, for the first time, truly planetary. Although, in the *Grundrisse*, Marx points out that the tendency to create the world market is intrinsic to the very concept of capital,[14] this potentiality was not fully actualized until very recently. Indeed, major Marxist theories of imperialism emerging in the context of the Second International and beyond were tacitly or explicitly premised on the assumption that capitalism was a relatively local phenomenon. Figures such as Luxemburg, Lenin, and Hilferding considered imperialism the means by which this local, Western organism (capital), interacted with its noncapitalist outside (non-Western

---

12   Moishe Postone, "Lukács and the Dialectical Critique of Capitalism," in Robert Albritton and John Simoulidis (eds.), *New Dialectics and Political Economy* (New York: Palgrave, 2003).

13   Achille Mbembe, *Critique of Black Reason* (Durham, NC: Duke University Press, 2017).

14   Karl Marx, *Grundrisse: Foundations of the Critique of Political Economy* (New York: Penguin, 1973 [1939]), 408.

peripheries) by means of military conquest, pillage, and colonial domination, basically in order to sustain its own process of "expanded reproduction." Recent transformations in the geographies of capitalist industrialization, however, warrant a reconceptualization of the scale at which this process operates. According to Cammack,

> The commonly quoted assertion that growth has been slower since 1973 than before is true for the West, and for the world as a whole. But Asia is the exception. Add to this the observation that the "world market" envisaged by Marx and Engels only a century and a half ago only came into being in the 1990s, and it is reasonable to suggest that far from it being the case that capitalism is a "Western" phenomenon, its "Western" phase, protracted though it seems from the point of view of the present, has merely coincided with its pre-history as a genuinely global form.[15]

The idea that capitalism has become emancipated from its traditional heartland and global in its geographical extent is epochal, and has brought back old debates on the "new international division of labor" (NIDL).[16] After being highly influential when originally formulated by Folker Fröbel, Jürgen Heinrichs, and Otto Kreye in 1977, the NIDL thesis fell out of favor in the late 1980s. In the face of declining profitability and increasing labor unrest, Fröbel et al. argued,[17] companies embarked on a process

---

15 Paul Cammack, "The Shape of Capitalism to Come." *Antipode* 41, no. S1 (2009): 265.

16 Ho-Fung Hung, "Introduction: The Three Transformations of Global Capitalism," in Ho-Fung Hung (ed.), *China and the Transformation of Global Capitalism* (Baltimore: Johns Hopkins University Press, 2009); Sandro Mezzadra and Brett Neilson, *Border as Method, or, the Multiplication of Labor* (Durham, NC: Duke University Press, 2013, chapter 3; Guido Starosta, "The Outsourcing of Manufacturing and the Rise of Giant Global Contractors: A Marxian Approach to Some Recent Transformations of Global Value Chains," *New Political Economy* 15, no. 4 (2010): 543–63; Greig Charnock and Guido Starosta (eds.), *The New International Division of Labour: Global Transformation and Uneven Development* (London: Palgrave McMillan, 2016); Ray Hudson, "Rising Powers and the Drivers of Uneven Global Development," *Area Development and Policy* 1, no. 3 (2016): 279–94; Jamie Peck, "Macroeconomic Geographies," *Area Development and Policy* 1, no. 3 (2016): 305–22.

17 Folker Fröbel, Jürgen Heinrichs, and Otto Kreye, *The New International Division of Labour: Structural Unemployment in Industrialised Countries and Industrialisation in Developing Countries* (Cambridge: Cambridge University Press, 1980).

of global organizational restructuring that involved relocating industrial production to "Third World" countries. Critics argued that there was nothing essentially new about the NIDL because, although many Third World countries were no longer exporting raw materials but manufactured goods, the processes that had been spatially relocated involved low-skill, labor-intensive tasks, so the hierarchical international system, structured around Western cores and their dependent peripheries and semiperipheries, remained intact. The startling industrial upgrading that began to take place after the 1990s, especially among first- and second-generation "Asian Tigers" (Japan, South Korea, Taiwan, Singapore), which later built up to the spectacular rise of China, put into question the geometries of power inherited from the classical world system.

Ever since the long sixteenth century, and as authors in the world-systems-analysis tradition show, historical cycles of accumulation have been underpinned by scientific-technological revolutions in the forces and relations of production that allow "hegemonic powers" to achieve trade dominance. However, Marx argues that it was not until mechanized industry enabled machinofacturing in nineteenth-century England, that the technological foundation of large-scale industry became fully realized and the modern mode of production was able to "stand on its own feet."[18] In *Late Capitalism*, Ernest Mandel recounts three major revolutions in the technology of the production of motive machines by machines:[19] machine production of steam-driven motors since 1848, machine production of electric and internal combustion systems since the 1890s, and machine production of nuclear-powered apparatuses since the 1940s.

One of the main contentions of this book is that the capacity for capital to radically upscale its metabolic process has been the result of a leap forward in the productive technology of large-scale industry enabled by a fourth machine age. This technological reconfiguration is not an exclusive outcome of the miniaturization and computerization of industrial systems brought about by recent innovations in semiconductor materials based on integrated circuits. Neither is it a product of advances in machine learning, digital fabrication, gene sequencing, or biotechnology. The fourth machine age emerges from the fusion of such technologies and their interaction across physical, digital, and biological domains, a

---

18   Marx, *Capital*, Vol. 1, 506.
19   Ernest Mandel, *Late Capitalism* (London: NLB, 1976 [1972]).

feature that qualitatively distinguishes it from previous industrial revolutions.[20] However, this book reclaims Mandel's aspiration to reject uncritical revisionism in the face of technological novelty. The emergence of a fourth stage or moment in the historical evolution of capitalism does not imply that there has been a break away from the intrinsic modes of motion of modern society—that is, the scientific-technological organization of human exertion through large-scale industry and the reproduction of the class antagonism under the aegis of the wage-system. Quite to the contrary, and as Fredric Jameson points out, Mandel employs the notion of late capitalism precisely "to demonstrate that it is, if anything, a purer stage of capitalism than any of the moments that preceded it."[21]

In the mining industry, the synergistic effect of innovations in robotics, biotechnology, artificial intelligence, and geospatial information systems, has exerted a fundamental overhaul in the extraction and processing of minerals. The implementation of such technologies, according to official figures, has allowed Latin American countries to achieve substantial increases in mineral exports relative to total exports—going, for example, from 39.7 percent in 2001 to 62.4 in 2010 in Chile; from 46 percent in 2000 to 65 percent in 2010 in Colombia; and from 45 percent to 61 percent for the same period in Peru.[22] The economic profitability brought about by the smart and robotized mine, however, pales in comparison to its material footprint, as an average large-scale extraction site produces up to a thousand times more solid

---

20   Jaap Bloem, Meno van Doorn, Sander Duivestein, David Excoffier, René Maas, and Erik van Ommeren, *The Fourth Industrial Revolution: Things to Tighten the Link Between IT and OT* (Groeningen: Sogeti VINT, 2014); Erik Brynjolfsson and Andrew McAfee, *The Second Machine Age: Work, Progress, and Prosperity in a Time of Brilliant Technologies* (New York: W.W. Norton, 2014); Andrew D. Maynard, "Navigating the Fourth Industrial Revolution," *Nature Nanotechnology* 10 (2015): 1005; Klaus Schwab, *The Fourth Industrial Revolution* (Geneva: World Economic Forum, 2016). For the interaction of such technologies in the mining industry, see Freyja Knapp, "The Birth of the Flexible Mine: Changing Geographies of Mining and the E-Waste Commodity Frontier," *Environment and Planning A* 48, no. 10 (2016): 1889–1909; Labban, "Deterritorializing Extraction."

21   Fredric Jameson, *Postmodernism, or, The Cultural Logic of Late Capitalism* (London and New York: Verso), 3.

22   Arturo Cancino, "La dudosa fortuna minera de Suramérica: los países andinos Colombia, Chile y Perú," in Catalina Toro Pérez, Julio Fierro Morales, Sergio Coronado Delgado, and Tatiana Roa Avendaño (eds.), *Minería, Territorio y Conflicto en Colombia* (Bogotá: Universidad Nacional de Colombia, 2012), 66.

waste than those working with older technologies.[23] To put this figure
into perspective, a large open-cast mine can produce up to forty times
more solid waste in one year than any Latin American megacity produces
during the same time period.[24] Considering the radical externalization
of biogeophysical costs that has developed in tandem with the increase
in the production of raw materials, the fourth technological revolution
has so far been unrevolutionary. For this reason, this book considers the
capital-intensive mine to be an offshoot of a fourth machine *age*, rather
than of a fourth technological *revolution*.[25]

Although anomalous and suboptimal when compared with previous
industrial eras, the quantum leaps in labor productivity enabled by the
fourth machine age have nonetheless brought about a geo-economic
shift in the process of global capital accumulation.[26] If the horizontal
resource frontiers of coal and rubber begotten by the first machine age
embodied the ascendancy of the British Empire, and the vertical frontiers of oil extraction of the second and third technological revolutions
elevated the United States to the status of indisputable superpower
during most of the twentieth century, the planetary mine encapsulates a
world-historical transformation of comparable significance. These
historically unique geographies of extraction, increasingly populated by
robots and scientists, have underpinned the process by which the Asian
Tigers went from being mere export-processing zones immersed in
"captive" positions within global supply chains to becoming industrial

23   Interview with a geophysicist from London Mining Network, September 11,
2015.

24   Mauricio Cabrera and Julio Fierro, "Implicaciones ambientales y sociales del
modelo extractivista en Colombia," in Garay Salamanca and Luis Jorge (eds.), *Minería
en Colombia: fundamentos para superar el modelo extractivista* (Bogotá: Contraloría
General de la República de Colombia, 2013).

25   I use the concept of "machine age" to anchor my interpretation of technological
modernization in the materialist periodization of large-scale industry Ernest Mandel
developed in his landmark work *Late Capitalism*. The difference between the concepts
of "machine age" and of "industrial revolution" are therefore not reducible to a matter of
semantics but imply a deliberate effort to depart from the circulationist approaches that
tend to predominate in the mainstream literature. The contradictions and pitfalls of the
concept of "the fourth industrial revolution," popularized by Schwab, *Fourth Industrial
Revolution*, and Brynjolfsson and McAfee, *Second Machine Age*, among other authors,
are fully elucidated in chapter 2.

26   Jason W. Moore, *Capitalism in the Web of Life: Ecology and the Accumulation of
Capital* (New York: Verso, 2015), chapter 6.

powerhouses and logistical juggernauts whose scale of industrial dynamism is without parallel in history.

The dispersed fragments of the planetary mine are therefore dialectically bound to the pipes and cables tucked into the high-rise buildings of the megacities that have sprung out of nowhere in south China in just twenty years;[27] in the myriad electronic gadgets spawned by vertically integrated electronics-manufacturing systems that have led to factories in Shenzhen employing up to 420,000 workers—a figure no automobile factory in the heyday of Fordism would have even dreamed of;[28] and in the hulls and containers of the thousands of cargo vessels that make up the maritime commercial fleets of China, Japan, and South Korea, which have rendered the Pacific Ocean the main infrastructural corridor of world trade.[29] Perhaps most importantly, the metabolism of the supply chain of extraction is also objectified in those unspectacular, nearly imperceptible practices and habits that constantly weave together the fabric of everyday life in the twenty-first-century city: sending an email, driving to work, ordering groceries through the internet. It is an intrinsic feature of the commodity form and the bourgeois ideologies and epistemological frameworks that support it to constantly and systematically break the totality of the urban experience into apparently disconnected and unhistorical fragments. A further objective of this book is therefore to overcome the fetish of the commodity and to make these manifold metabolic mediations visible and conceptually intelligible. In this sense, the book intends to contribute to works that have considered extraction to involve much more than the mere spatiality of the mine,[30]

27 Jia Ching Cheng, "Hukou: Labor, Property, and Urban-Rural Inequalities," in Stefan Al (ed.), *Factory Towns of South China: An Illustrated Guidebook* (Hong Kong: Hong Kong University Press, 2012), 37; Wade Shepard, *Ghost Cities of China* (London: Zed Books, 2015).

28 Stefan Al (ed.), *Factory Towns of South China*; Boy Lüthje, Stefanie Hürtgen, Peter Pawlicki, and Martina Sproll, *From Silicon Valley to Shenzhen: Global Production and Work in the IT Industry* (Lanham, MD: Rowman and Littlefield, 2013); Brian Merchant, *The One Device: A Secret History of the iPhone* (Boston: Little, Brown, 2017).

29 Paul Ciccantell, "China's Economic Ascent and Japan's Raw-Material Peripheries," in Ho-Fung Hung (ed.), *China and the Transformation of Global Capitalism*; Parag Khanna, *Connectography: Mapping the Global Network Revolution* (London: Weidenfeld & Nicholson, 2016); Organization for Economic Cooperation and Development (OECD), "Logistics Observatory for Chile: Strengthening Policies for Competitiveness," International Transport Forum, 2016.

30 Stephen Bunker and Paul Ciccantell, *Globalization and the Race for Resources*,

as well as to broader dialogues on urban political ecology and planetary urbanization.[31]

Although the book is driven by the aspiration to develop theoretical frameworks and political imaginations that can shed light into the making of spaces of extraction as capital escalates into a global (not merely Anglo-European) socionatural system, it is strongly grounded in empirical observation. It draws upon several years of research on the political economies of extraction in both Chile and Latin America broadly considered. The changing configurations of resource frontiers in Latin America illustrate the gravitational force exerted by the new geographies of industrialization and urban expansion assembled around the Pacific Ocean. The balance of trade between China and Latin America has expanded dramatically in recent years, going from $15 billion in 2009 to a staggering $200 billion in 2011.[32] According to the *Atlas of Economic Complexity*, Chile's exports of metals to China showed an astonishing increase between 2001 and 2011, going from $460 million to $11.1 billion.[33] It is quite telling that such a dramatic increase in the

---

Baltimore: Johns Hopkins University Press, 2005; Gavin Bridge, "The Hole World: Spaces and Scales of Extraction," *New Geographies* 2 (2009): 43–48; Labban, "Deterritorializing Extraction"; Mimi Sheller, *Aluminium Dreams: The Making of Light Modernity* (Cambridge, MA: MIT Press, 2014); Verónica Gago and Sandro Mezzadra, "A Critique of the Extractive Operations of Capital: Toward an Expanded Conception of Extractivism," *Rethinking Marxism* 29, no. 4 (2017): 574–91; Sandro Mezzadra and Brett Neilson, "On the Multiple Frontiers of Extraction: Excavating Contemporary Capitalism," *Cultural Studies* 31, nos. 2–3 (2017): 185–204; Felipe Irarrázaval and Beatriz Bustos-Gallardo, "Global Salmon Networks: Unpacking Ecological Contradictions at the Production Stage," *Economic Geography*, October 9, 2018.

31   Nik Heynen, Maria Kaika, and Erik Swyngedouw (eds.), *In the Nature of Cities: Urban Political Ecology and the Nature of Urban Metabolisms* (New York: Routledge, 2006); Neil Brenner, "Introduction: Urban Theory without an Outside," in Neil Brenner (ed.), *Implosions/Explosions: Towards a Study of Planetary Urbanization* (Berlin: Jovis, 2014); Erik Swyngedouw, *Liquid Power: Contested Hydro-Modernities in Twentieth-Century Spain* (Cambridge, MA: MIT Press, 2015); Matthew Gandy, *The Fabric of Space: Water, Modernity, and the Urban Imagination* (Cambridge, MA: MIT Press, 2017); Roger Keil, *Suburban Planet: Making the World Urban from the Outside* (Cambridge, UK: Polity, 2018).

32   Rafael Valdez Mingramm, "China y América Latina hacia el 2030, colaboración estratégica y colaboración energética," in Trápaga Delfín, Yolanda (ed.), *América Latina y el Caribe-China: Recursos Naturales y Medio Ambiente*, México DF: Unión de Universidades de América Latina y el Caribe, 2013, 32.

33   Chile's exports to China during the 1994–2014 period, from the *Atlas of Economic Complexity*, atlas.cid.harvard.edu/explore/stacked/export/chl/chn/show/1995.2014.2, accessed April 19, 2017.

volume of exports began after 2001, a year that also saw the beginning of the aggressive movement toward mechanization in the mining industry depicted in figure 1. Such figures, however, do not imply that China has become the new imperial or "hegemonic" power of the global economy, in a manner akin to those that mark the systemic cycles of accumulation once theorized by Giovanni Arrighi.[34] The processes that shape contemporary geographies of extraction, as we will see in the next section, warrant a thorough rethinking of the very nature of capitalist imperialism, political authority, and the spatial architecture of the world market.

Excavating the Planetary

Recent efforts to grapple with the notion of the planetary have come from variegated intellectual traditions, politico-normative orientations, and epistemological standpoints.[35] However, they all share the assumption that the world depicted by major theories of globalization has been rendered unrecognizable by interconnected crises of liberal cosmopolitanism, of global warming, and of encroaching concentration of wealth. The very idea of the "global" as the proverbial blue marble demarcated and measured through grids and coordinates is being gradually superseded by that of the "planetary," in which the earth reemerges as an unfamiliar place riddled with eerie, destructive, and menacing forces.[36] As opposed to the "spaces of flows" and "liquid

---

34 Giovanni Arrighi, *The Long Twentieth Century: Money, Power and the Origins of our Times* (New York: Verso, 2010 [1994]).

35 Gayatri Chakraborty Spivak, *Death of a Discipline* (New York: Columbia University Press, 2003); James D. Sidaway, Chih Yuan Woon, and Jane M. Jacobs, "Planetary Postcolonialism," *Singapore Journal of Tropical Geography* 35 (2014): 4–21; Brenner (ed.), *Implosions/Explosions*; William E. Connolly, *Facing the Planetary: Entangled Humanism and the Politics of Swarming* (Durham, NC: Duke University Press, 2017); Bruno Latour, *Facing Gaia: Eight Lectures on the New Climatic Regime* (Cambridge: Polity, 2017); Dipesh Chakrabarty, "Planetary Crises and the Difficulty of Being Modern," *Millennium* 46, no. 3 (2018): 259–82; Silvia Rivera Cusicanqui, *Un mundo ch'ixi es posible: ensayos desde un presente en crisis* (Buenos Aires: Tinta Limón Editores, 2018); Geoff Mann and Joel Wainwright, *Climate Leviathan: A Political Theory of Our Planetary Future* (New York: Verso, 2018).

36 Lefebvre hints at some of the preliminary manifestations of this shift by claiming that the world market was outlining new sociospatial configurations that were becoming

modernities" that populated earlier visions of globalization, the notion of the planetary designates a convoluted terrain where fences, walls, and militarized borders coexist with sprawling supply chains and complex infrastructures of connectivity. This realm is traversed by deeply contradictory and yet complementary tendencies toward advanced functional integration in the world economy and toward radical ethnoracial and sociospatial fragmentation. Crucially, the shift from the global to the planetary is also understood as a steppingstone toward novel formations of collective consciousness and of collective agency.

The planetary mine is the geography of extraction that emerges from and underpins this contradictory state of things. If we put together all the phases that comprise the transnational circulatory system of primary-commodity production and trace the journey of copper from its point of extraction in the Andean plateaus of Chile to its point of destination in the spaces of advanced manufacturing in China, a bewildering panorama emerges: Autonomous trucks and shovels working at nearly 4,000 meters above sea level put the metal into a semiautomated train, which then takes it to a smelting and electrorefining facility, where computerized ovens transform it into copper cathodes. The cathodes are put into containers and sent to one of the megaports of the mining industry in the Atacama Desert, where gantry cranes load the cargo into a container ship. After crossing the Pacific, our container is unloaded by the swiftness of the vast mechanical systems of the capital-intensive Chinese ports. Finally, the copper cathodes end up in one of the infamous "dark factories" of the Pearl River Delta. Here, robots and computer numerical control (CNC) machine tools operate in the dark, turning copper into the wires that hundreds of thousands of human laborers in electronics assembling facilities will later etch into the electronic gadgets we carry in our pockets.

The above is, of course, a stylized and oversimplified description, but it illustrates how the commodity relation reconfigures the social

---

inscribed on the terrestrial surface of changing spaces, doing away with inherited geopolitical units. The emergence of the planetary, according to Lefebvre, challenged the epistemological logic of traditional categories of political economy. Henri Lefebvre, "The Worldwide and the Planetary," in Neil Brenner and Stuart Elden (eds.), *Henri Lefebvre: State, Space, World*, Minneapolis: University of Minnesota Press, 2009 [1973].

character of labor into a hostile and uncanny system of alien objectivity that turns against the very peoples and ecologies who produce it. Moishe Postone's 1993 book, *Time, Labor, and Social Domination*, a harbinger of the present state of things, foreshadowed the aggressive movement toward advanced automation and industrial expansion that would begin to take place in the postglobalization context. This landmark work suggested that the accumulated and collective species-powers of humanity were being expropriated and remodeled into the attributes of an alienated system of social domination that exists objectively and quasi-independently in machines, algorithms, and individual labors. In this sense, Postone considered that, as a directionally purposed form of social mediation, capital needed to be understood as an entity of an entirely different ontological order than a mere thing or process. To the extent that capital displays attributes traditionally considered uniquely human, such as intentionality (manifested in the endless drive to valorize itself), and also becomes actively embodied in objects and individuals, Postone considered that it was most adequately understood as an "alienated subject" (ibid). This insight had been partially developed by Marx, who described the more-than-human scripture of power enacted by the movement of value:

> In the circulation M-C-M both the money and the commodity function only as different modes of existence of value itself, the money as its general mode of existence, the commodity as its particular or, so to speak, disguised mode. It is constantly changing from one form into the other without becoming lost in this movement; *it thus becomes transformed into an automatic subject* . . . For the movement in the course of which it adds surplus-value is its own movement, its valorization is therefore self-valorization [*Selbstverwertung*]. By virtue of being value, it has acquired the occult ability to add value to itself. [emphasis added][37]

The significance of Postone's contribution consists in the systematic theory of neoliberal globalization he developed based on this insight, which was already latent in the Marxian critique. In its materialist reinterpretation of Hegel's category of Spirit or *Geist*—the identical,

---

37   Marx, *Capital*, Vol. 1, 255.

world-historical subject—*Time, Labor, and Social Domination* might be as important for the twenty-first century as Lukács's *History and Class Consciousness* was for the twentieth. However, whereas Lukács equates the proletariat with the identical subject-object of the historical process, Postone argues that it is actually capital that displays the features of the objectifying, world-making practice of the *Geist*. As the *Geist*, however, Postone considers capital to be a remarkable subject because—whereas Hegel's subject was transhistorical, ideal, and knowing—capital is historically determinate, quasi-organic, and blind.[38] In its alienated, spiraling movement of self-expansion, capital becomes constitutive of specific forms of social practice and subjectivity. These forms of practice, Postone highlights, subject people, things, and institutions to impersonal modalities of social domination.

It is not difficult to see how money, commodities, technical artifacts, labor, and the state—when determined as modes of existence of capital—have given rise, to paraphrase Marx, to an "enchanted and perverted world" that appears to be demonically out of joint. In the mining industry, as the following chapters will explore in detail, the self-objectifying practice of capital has metamorphosed into rivers poisoned by mercury and cyanide, ancient glaciers torn to shreds by explosives and machineries of extraction, peasants ravaged by debt, police forces bizarrely out of control, mining towns riddled by cancer epidemics, and rampant labor casualization. In rendering visible how human bodies become possessed (and often obliterated) by uncanny forces and nonhuman objects become animated with powers over life and death, this book unfolds a subversive reading of the planetary: one that takes seriously the class-based, contradictory, and inherently antagonistic dimensions of contemporary sociotechnical systems and ecologies and points toward the possibilities the current industrial era brings for collectively imagining and producing a different kind of modernity.

Understanding the mine as an interconnected system of spatial technologies and infrastructures allows for superseding the superficial appearance of spaces of extraction as exclusively fragmented and sclerotic. Such a shift of vision is crucial for the core objective of this book,

---

38   Postone, "Lukács and the Dialectical Critique of Capitalism."

which is to problematize the methodological nationalism that, to this day, informs major accounts of resource extraction. Informed by theories of dependency, world-systems analysis, structuralist economics, state regulation theory, and even certain variants of neoclassical economics, the vast majority of studies of natural resources have considered the dynamics of primary-commodity production to be determined by the international political relations of the nation-state, whether in the guise of dependency, imperialism, and/or unequal exchange between cores and peripheries. Although those dynamics persist, I argue that they are more adequately conceptualized as the fetishized form of expression of a process whose content transcends an interstate system of political actors with apparent autonomy from the class antagonism that lies at the heart of the historical movement of modern society. Roughly put, the shift from the global to the planetary entails a process of "unthinking"—to use Immanuel Wallerstein's apt formulation—whereby the existing conceptuality of society is subjected to critical inquiry on the basis of the practical, real relations of social life—not the other way around.

The idea of the "national economy" as an imagined community of relatively homogeneous interests, and therefore to some extent politically autonomous from the class struggle, has always been problematic. In the early years of the twentieth century, a young and brilliant Rosa Luxemburg exposed the antinomies of this concept as she addressed the Socialist International in magnetic words: "Why speak of 'national self-determination'? Under capitalism the nation does not exist! Instead we have classes with antagonistic interests and rights. The ruling class and the enlightened proletariat can never form an undifferentiated national whole."[39] Although illusory, the political body that is the modern state form is nevertheless objectively real (just as a container ship is "real" despite being a form of existence of an abstract process) and constitutes the medium through which the world market unfolds its potentialities. This means that the political authority of the former is the precondition for the very existence of the latter.[40] The two constitute a contradictory

---

39   Rosa Luxemburg, cited in Kate Evans, *Red Rosa* (New York: Verso, 2015), 41.

40   Claudia von Braunmühl, "On the Analysis of the Bourgeois Nation State within the World Market Context: An Attempt to Develop a Methodological and Theoretical Approach," in John Holloway and Sol Picciotto (eds.), *State and Capital: A Marxist Debate* (Austin: University of Texas Press, 1978); Werner Bonefeld, "Social Constitution and the

yet unitary political-economic logic. The world market, on such basis, should therefore not be seen as a sum of national economies but as a sociomaterial system organized in the form of national economies as its aliquot parts. This methodological principle, where the totality takes prevalence over its component elements, allows for making sense of the contradiction intrinsic to the very idea of planetarity: the fact that as the world market expands across the globe, the late liberal state becomes increasingly militaristic, interventionist, and coercive.[41]

The immanent content that governs the production and reproduction of spaces of extraction under late capitalism, this book therefore argues, is the tendency to increase the organic composition of capital, and the concomitant fragmentation of the productive subjectivity of the international working class. This insight takes inspiration from Marxian readings of the new international division of labor.[42] Insisting on the problem of class fragmentation is a pressing matter in both intellectual and political terms. Amid the unprecedented material interdependence of concrete labors spawned by advanced industrialization, individuals have been separated from each other by ideology and are therefore unable to realize that despite considerations of nationality, culture, and

---

Form of the Capitalist State," in Bonefeld, Gunn, and Kosmas (eds.), *Open Marxism*; Iñigo Carrera, *El Capital*; David McNally, "The Blood of the Commonwealth: War, the State, and the Making of World Money," *Historical Materialism* 22, no. 2 (2014): 3–32.

41   The notion of totality is mobilized here to foreground the radical interdependence of social and ecological existence under capitalism. As such, it departs from the misinterpretations of this concept that are usually found in some traditions of poststructuralist thought and postcolonial studies. In such accounts, totality or totalization tend to be understood as a shorthand for homogenization—or, even more strikingly, for totalitarianism. For a thorough discussion on such misrepresentations, see Kanishka Goonewardena, "Planetary Urbanization and Totality," *Environment and Planning D: Society and Space* 36, no. 3 (2018): 456–73; Vivek Chibber, *Postcolonial Theory and the Specter of Capital* (New York: Verso, 2013). Chapter 6 addresses in depth the question of totality through an exploration of the scales and levels of social reality in which the financial system becomes intermingled with primary-commodity production.

42   Gary Nigel Howe, "Dependency Theory, Imperialism, and the Production of Relative Surplus Value on a World Scale," *Latin American Perspectives* 8, no. 3/4 (1981): 82–102; Maria Mies, *Patriarchy and Accumulation on a World Scale: Women in the International Division of Labor* (London: Zed Books, 2014 [1986]); Juan Iñigo Carrera, "The General Rate of Profit and Its Realisation in the Differentiation of Industrial Capitals," Guido Starosta, "Revisiting the New International Division of Labour Thesis," and Gastón Caligaris, "The Global Accumulation of Capital and Ground-Rent in "Resource-Rich" Countries," in Charnock and Starosta (eds.), *The New International Division of Labour*.

ethnicity, they share increasingly common conditions of existence. In a 1954 poem titled "Ode to Copper," Pablo Neruda's pen captures with might and clarity of vision the transcontinental, dialectical, yet opaque relations of social interdependence engendered by resource extraction. In this poem, Neruda asks a miner, his "brother of the rooted summit,"

> for this did you give it [copper] birth amid the pains? So it became a menacing cyclone, tempestuous misfortune? So it demolished the lives of the poor, the other poor, your own family that perhaps you are unacquainted with, and spreads across the world?[43]

Throughout the commodity supercycle of the turn of the century, millions upon millions of indigenous and *campesino* communities have lost their livelihoods in Latin America as infrastructures of extraction expand aggressively and destructively across the region's erstwhile countrysides. Many have been proletarianized or forced to migrate to the *favelas*, *villas miseria*, *comunas*, and *campamentos*, among some of the modalities of shantytowns in which Latin American cities have become ever more ensnared. However, is the plight of these peasants substantially different from that of the millions of Chinese migrant workers who have had to leave their families to work in the overcrowded, fractured, and polluted manufacturing cities of the *hukou* system? Or that of the subcontracted workers in the logistics warehouses of the United States, whose children go to bed on an empty stomach most nights? Even in the preliminary phases of the new international division of labor in 1986, Maria Mies detected how the universalizing powers of patriarchal capitalism led to the paradoxical state of affairs in which "Third World" and "First World" women are at once materially connected and subjectively fragmented by commodity production.[44] The former, exploited in factories as producers, cannot exist without the latter subjugated in households as consumers. Moreover, Mies considered that the standard of living of the two was becoming increasingly aligned, as deindustrialization and neoliberalization in the global North was giving rise to a "new poor" whose conditions of existence gradually came to resemble those of women in the so-called underdeveloped countries,

---

43   Pablo Neruda, *All the Odes* (New York: Farrar Straus Giroux, 2013), 171.
44   Mies, *Patriarchy and Accumulation*.

where large-scale industry had been offshored. These two actors, however, have each remained completely oblivious to the existence of the other.

At the heart of *Planetary Mine* is, therefore, an attempt to lay bare the networks of relations that connect the lives of the workers and local communities in spaces of extraction to those so far considered external to the worlds of primary-commodity production—such as migrants, technocrats, financiers, and workers toiling in ports, warehouses, and factories, among others. With this, I argue that the polarizing tendencies of uneven geographical development are no longer exclusively confined to the international political relations of the nation-state (specifically the rift between manufacturing centers and resource-rich countries) but have become immanent to the production of space under capitalist society. The periphery needs to be understood as a ubiquitous sociospatial condition, not as the exclusive domain of international political relations. This in no way implies that the state has been rendered irrelevant to the making of primary-commodity frontiers or that its sovereignty has become eroded by the insurmountable forces of the world market. Considering global capitalism to be a faceless, "structural" force that exists in a separate plane of reality to that of the mundane workings of states and firms is politically paralyzing and analytically obfuscating. As authors in the tradition of *post-extractivismo* have shown, the messy materialities and entanglements of Latin American states have been crucial to the production and reproduction of geographies of extraction in the context of the commodity supercycle.[45] However, existing studies tend to mistake the state-*form* for its capitalist *content*; the class antagonism that underlies the very production of primary-commodity frontiers, whose existence is essentially international, is usually ignored or pushed into the background. For this reason, this book is also

---

45   Eduardo Gudynas, "Diez tesis urgentes sobre el nuevo extractivismo," in Centro Latinoamericano de Ecología Social (ed.), *Extractivismo, Política y Sociedad* (Quito: Centro Andino de Acción Popular, 2009); Maristella Svampa and Marian Sola Álvarez, "Modelo minero, resistencias sociales y modelo de desarrollo: los marcos de la discusión en la Argentina," *Ecuador Debate/Tema Central* (2010): 105–26; Catalina Toro, "Geopolítica energética: minería, territorio y resistencias sociales," in Toro, Fierro, Coronado, and Roa (eds.), *Minería, Territorio y Conflicto en Colombia* (Bogotá: Universidad Nacional de Colombia, 2012).

concerned with elucidating the properly global content of resource imperialism and of the bourgeois state-form in the context of contemporary geographies of extraction.

On Resource Imperialism and Revolutionary Subjectivity

Though the theory of imperialism was the crown jewel of Marxist thought for decades, Panitch and Gindin point out how it has recently become an impracticable labyrinth where no one seems able to find their way.[46] Indeed, with the demise of formal colonization processes after the mid-twentieth century, the traditional reading of resource peripheries as the embodied expression of relations of imperial domination became problematic, to say the least. One of the most relevant contributions of authors in the tradition of world-systems analysis, in my view, has been to lay bare the relations between resource frontiers and the empire-making projects that have shaped the capitalist world economy since the long sixteenth century.[47] After all, it was the gold and silver coming from the mines of the Americas that supported the Habsburg monarchies in the Iberian Peninsula and the monetary system articulated around the Italian city-states of Venice and Genoa. Interoceanic trade of timber and peat allowed the Dutch Empire to become the financial powerhouse of Europe in the seventeenth century, just as South American guano and rubber would later feed the farms and factories of the sprawling British Empire. The vertical oil frontiers of the twentieth century, in turn, enabled the United States to assemble the largest and most sophisticated military apparatus the world has ever known, as well as to introduce a universalizing model of capitalist urbanization centered around the automobile.

The rise of China, however, complicates a state-centric reading of the capitalist world system in significant ways. To begin with, the internationalization of the Chinese state and that of its region of influence in no

---

46   Leo Panitch, Sam Gindin, "Global Capitalism and American Empire," *Socialist Register* 40 (2014): 1–42.

47   Arrighi, *Long Twentieth Century*; Bunker and Ciccantell, *Globalization*; Moore, *Capitalism and the Web of Life*.

way resembles anything close to that of the "hegemons" that led the
systemic cycles of accumulation in Arrighi's landmark periodization.[48]
According to Parag Khanna, China has built a "global supply chain
empire" that extends outward to myriad resource peripheries without
fighting a single skirmish.[49] China's global presence has thus been
defined not by its military or even its political emissaries, but by its
supply chains. China's strategy, Khanna illustrates, has not been to
occupy countries formally or even to intervene in their domestic political landscape, but to ease passage across them. Carmody and Taylor
have argued that China's strategies for securing access to raw materials
abroad are better viewed as a result of the novel framework of horizontal, cooperative, noninterventionist international relations that they
refer to as *flexigemony*.[50] In Latin America, the consensus among scholars of international relations is that the rise of China marks a break with
the vertical, militaristic, and interventionist stance through which
Western economic powers traditionally acquired access to natural
resources.[51]

---

48   Arrighi, *Long Twentieth Century*.
49   Khanna, *Connectography*. Although it is usually argued that China's role in the
interstate system resembles the cosmopolitan-mercantile orientation of the Dutch
Empire, Giovanni Arrighi and Jason W. Moore have warned against the misleading
character of such extrapolations: "Capitalist Development in World Historical
Perspective," in Robert Albritton, Makoto Itoh, Richard Westra, and Alan Zuege (eds.),
Phases of Capitalist Development: Booms, Crises, and Globalizations (London: Palgrave
McMillan, 2001; Giovanni Arrighi, *Adam Smith in Beijing: Lineages of the Twenty-First
Century* (New York: Verso, 2009). For these authors, the rise of Asian Tigers constitutes
a double anomaly when set against the background of previous systemic cycles of
accumulation. First, the world's emerging creditor nations (Japan, Taiwan, Hong Kong,
Singapore, and most recently mainland China) signal a reversal from the shift of
financial power from the non-Western to the Western world. Second, and most
importantly, the military and economic powers that constitute these emerging
hegemonic forces are not centralized in a nation-state, or even several of them. The
China-centered regional world system, Arrighi and Moore suggest, is organized
politically in city-states (one sovereign, Singapore, and one semisovereign, Hong Kong),
a semisovereign province (Taiwan), and a military protectorate of the United States
(Japan). New generations of Asian Tigers (often referred to as "Tiger Cub economies"),
such as Malaysia, Vietnam, Indonesia, the Philippines, and Thailand, further complicate
straightforward comparisons with the cycle of accumulation centered around the Dutch
Empire.
50   Pádraig Carmody, Ian Taylor, "Flexigemony and Force in China's Resource
Diplomacy in China: Sudan and Zambia Compared," *Geopolitics* 15 (2010): 496–515.
51   Rubén González-Vicente, "Mapping Chinese Mining Investment in Latin

The material articulation of a global sphere of accumulation would then appear to have banished imperialism from the face of the earth. Appearances, however, can be misleading. If the politics of empire seems to be an anachronism that belongs to previous stages of capitalist development, it is perplexing that the local communities and workers who coexist with geographies of extraction tend to experience the expansion of primary-commodity frontiers systematically in idioms of imperialism and neocolonial domination.[52] In this sense, the violent dislocations spurred by the world market are much more than the result of a mere economic mediation, and this fact emerges with full clarity in the minds of the oppressed. As Holloway forcefully argues, the apparent fragmentation of the mystifying forms of modern society come into constant conflict with the workers' total experience of class oppression.[53] The machinery, police trucks, and private security personnel—often working in connivance with death squads—swarming around extraction sites, in this sense, are not experienced by indigenous communities as part of a postimperialist context of harmonious international relations. They are, rather, seen as the material embodiment of a more advanced stage of the same system of capitalist domination sedimented in centuries of racialized violence, enslavement, and expulsion. For Deborah Cowen, it is therefore both striking and diagnostic that old enemies of empire—"Indians" and "pirates"—are

America: Politics or Market?," *China Quarterly* 209 (2012): 35–58; Yolanda Trápaga Delfín, "Introducción," in Yolanda Trápaga Delfín (ed.), *América Latina y el Caribe-China: Recursos Naturales y Medio Ambiente* (Mexico City: Unión de Universidades de América Latina y el Caribe, 2013); Loretta Napoleoni, "La política china de ventajas mutuas," in Enrique Dussel Peters (ed.), *América Latina y el Caribe-China: Economía, Comercio e Inversiones* (Mexico City: Unión de Universidades de América Latina y el Caribe, 2013); Adriana Roldán Pérez, Alma Sofía Castro Lara, Camilo Alberto Pérez Restrepo, Pablo Echavarría Toro, and Robert E. Ellis, *La Presencia de China en América Latina: Comercio, Inversión y Cooperación Económica* (Medellín: EAFIT, 2016).

52   Aviva Chomsky and Steve Striffler, "Empire, Labor, and Environment: Coal Mining and Anti-Capitalist Environmentalism in the Americas," *International Labor and Working Class History* 85 (Spring 2014): 194–200; Aviva Chomsky, "Empire, Nature, and the Labor of Coal: Colombia in the Twenty-First Century," *Labor* 13, nos. 3–4 (2016): 197–222; Todd Gordon and Jeffery R. Webber, *Blood of Extraction: Canadian Imperialism in Latin America* (Black Point, NS: Fernwood Publishing, 2016); Jeffery R. Webber, "Contemporary Latin American Inequality: Class Struggle, Decolonization, and the Limits of Liberal Citizenship," *Latin American Research Review* 52, no. 2 (2017): 281–99.

53   John Holloway, "The State and Everyday Struggle," in Clarke (ed.), *State Debate*.

among the groups that pose the biggest threats to the "security" and homeostasis of global supply chains for the production and circulation of commodities.[54]

In light of the above considerations, this book proposes to under-stand imperialism as *one of the phenomenal forms in which global value relations assert themselves*. This means that capitalist imperialism—as opposed to dominant readings—is not autonomously determined by the political relations of the nation-state but by the directionally purposed drive to increase the organic composition of capital at the system-wide level. Inspired by the experience of struggle against large-scale mining in Latin America, but also in line with recent discussions on the NIDL,[55] *Planetary Mine* argues that the immanent dynamics that underpin the spaces of extraction of late capitalism are *global in content and national only in form*. This means that the process of capital accu-mulation on a world scale can often assume the distorted political forms of an imperialist or subimperialist intervention. Like all forms of fetish-ism, such forms are essentially illusory, despite their concrete material existence. The content of these mediations, then, is not the pursuit for "hegemonic status" within the interstate system. Rather, it is the produc-tion of a technological exoskeleton that extends to multiple resource frontiers—in the form of railways, ports, waterways, power plants, roads, debt instruments, digital platforms, and so forth—in order to support the expanded reproduction of capital on the basis of machino-facturing. Contrary to Hardt and Negri's influential theorization,[56] I argue that the radical upgrading and upscaling of large-scale industry, not its "dematerialization," form the basis of capitalist power in the twenty-first century.

Seen in this light, this book intends to develop a theory of resource imperialism that, following Claudia von Braunmühl's lucid provocation, takes the world market—rather than the nation-state—as the *a priori* level of analysis.[57] Ellen Meiksins Wood's notion of "the empire of capi-tal," for example, adequately captures the evolution of imperialism from

---

54   Deborah Cowen, *The Deadly Life of Logistics: Mapping Violence in Global Trade* (Minneapolis: University of Minnesota Press, 2014).

55   Charnock and Starosta (eds.), *New International Division of Labour*.

56   Michael Hardt and Antonio Negri, *Empire* (Cambridge, MA: Harvard University Press, 2001).

57   Braunmühl, "On the Analysis of the Bourgeois Nation State."

a political mediation driven by the direct use of force to one underpinned by impersonal and economic compulsions.[58] However, Wood's theorization is, in the end, unable to supersede the methodological nationalism of traditional approaches, because she considers that the *raison d'être* of the empire of capital is the transfer of wealth from poor to rich nations and not the extension of the discipline of capital to living labor. According to Vivek Chibber, one of the most remarkable aspects of the theories of imperialism that emerged from the Second International was that they exercised a gravitational pull away from political theories of imperialism and toward systemic and economic ones.[59] The downside, however, is that they resulted in a deterministic, or "vulgar materialist," view that considered the state a mere instrument of the capitalist classes.[60] For this reason, the notion of imperialism developed in this book is complemented by two chapters entirely devoted to elucidating the status of the nation-state as the fetishized form of expression—rather than instrument or overseer—of the global unfolding of value.

Last but not least, this book is also concerned with excavating the social determinations of the revolutionary subjectivity of the laboring classes and the subaltern groups whose lives are being transformed by the ebb and flow of the mining supply chain. The determination of capital as the alienated subject of modern existence, according to Postone, gives rise to a state of things that would seem to deny the history-making practice of humans.[61] However, in fostering unprecedented material interdependence, the mode of universality assembled by capital opens the possibility for individuals to appropriate democratically what they had constituted collectively in alienated form. The sociotechnical foundation of large-scale industry, Postone argues, creates the basis for another mode of universality, one based on concrete specificity rather than on abstraction. Contemporary indigenous struggles in Latin America already hint at the possibility of another kind of generalized interdependence between peoples, ecologies, and technologies. Notions

---

58 Ellen Meiksins Wood, *Empire of Capital* (New York: Verso, 2005).
59 Vivek Chibber, "The Return of Imperialism to Social Science," *European Journal of Sociology* 45, no. 3 (2004): 427–41.
60 Clarke (ed.), *State Debate*.
61 Postone, *Time, Labor, and Social Domination*.

of *food sovereignty*,[62] of *the universal ayllu*,[63] of *the pluriverse*,[64] and of *ch'ixi modernity*,[65] to cite a few examples, have begun to envision a "third space" where the disparate domains of the local and the planetary can become interwoven—in dynamic, intricate, and often antagonistic ways—but without ever becoming hybridized or fused.

The leap forward undergone by the instruments of production of resource extraction, successive chapters show, has revolutionized the vital potentialities and productive attributes of the workers directly employed by the mining industry and their contractors. Indigenous and peasant communities have likewise been transformed by the frenzied movement of these infrastructural systems, becoming an arrogant oppositional force that invents new modes of inhabiting the world. Indigenous, *campesino*, migrant, and debtor groups, among other "plebeian" and "motley" fragments of the global collective labor force, are the ones who have been pushing forward the frontier of political possibility in recent years. We should take them seriously and try to learn from their struggles, dreams, and aspirations. In this sense, the book also attempts to unfold a reinterpretation of the working class as a revolutionary subject, doing away with teleological and Eurocentric readings where subaltern struggles are deemed inferior to those of the (white) industrial proletariat. However, this should in no way be interpreted as a naïve fetishization of the subaltern subject or of its precapitalist, culturally specific elements. Revolutionary coalitions, as David Graeber points out, have tended to hinge upon an alliance between the less alienated and the most oppressed organs of the social body.[66] In the case of primary-commodity production, it is the actual encounter between indigenous, *campesino*, and women's groups with scientists, artists, and engineers—itself an encounter between vernacular science and modern science—that has triggered some of the most vibrant, hopeful political forces in contemporary Latin America.

62   La Vía Campesina, *Struggles of La Vía Campesina for Agrarian Reform and the Defense of Life, Land, and Territories* (Waterfalls, Zimbabwe: La Vía Campesina, 2017).

63   Álvaro García Linera, *Forma valor y forma comunidad: aproximación teórica-abstracta a los fundamentos civilizatorios que preceden el ayllu universal* (Buenos Aires: Muela del Diablo, 2009 [1995]).

64   Arturo Escobar, *Designs for the Pluriverse: Radical Interdependence and the Making of Worlds* (Durham, NC: Duke University Press, 2018.

65   Rivera Cusicanqui, *Un mundo ch'ixi es posible*.

66   David Graeber, "The New Anarchists," *New Left Review* 13, 2003, 61–73.

The Division of Labor in *Planetary Mine*

In developing conceptual and methodological tools for making sense of the global geography that shapes spaces of extraction in the twenty-first century, this book covers a wide range of intellectual traditions, historical accounts, and key events in the social history of both Chile and Latin America broadly considered. I in no way intend to offer a definitive statement, but this book should be understood as an invitation to transcend the territoriality of the mine and of the national economic space where it is located. In this sense, considering the manifold themes that intersect with resource extraction as well as the vast variety of geographies of extraction across different parts of the world, there are relevant but also necessary limitations to this work. To begin with, the book centers on the role of Latin America within the new international division of labor. This means that, although some parallels are drawn with other "resource-rich" regions such as Africa and the Middle East, it remains systematically bound to the Latin American context and its relation with the ascent of Asian economies. Moreover, although war and armed conflict figure prominently in the making of resource frontiers,[67] this book only explores "low-intensity warfare," especially as manifested in police and state violence. The labor-intensive and violent ecologies of illegal mining, usually around "conflict minerals" such as coltan, cassiterite, and tungsten, are outside the purview of this book.

Because the materialist theory of the state developed in *Planetary Mine* is inspired mainly by the experience of struggle against Chile's expanding resource frontiers, it is also important to highlight that there are no systematic engagements with the cases of the postneoliberal "Pink Tide" governments of Latin America (such as those of Bolivia, Ecuador, and Venezuela). The contradictions and antinomies of the latter political projects—which illustrate the materialist theory of the state that this book proposes—have already been explored in depth by authors working from diverse traditions.[68] Throughout the book, I draw

67 Michael Klare, *Resource Wars: The New Landscape of Global Conflict* (New York: Henry Holt, 2002); Philippe Le Billon, *Wars of Plunder: Conflicts, Profits and the Politics of Resources* (Oxford: Oxford University Press, 2014).

68 Eduardo Gudynas, "Agropecuaria y nuevo extractivismo bajo los gobiernos progresistas de América del Sur," *Territorios* 5 (2010): 37–54; Diego Andreucci and Isabella Radhuber, "Limits to 'Counter-Neoliberal' Reform: Mining Expansion and the

examples from the postneoliberal governments associated with the Pink Tide, mainly as a means to better illustrate the more manifestly neoliberal and neodevelopmentalist regimes of natural-resource governance that predominate across most of Latin America and other world regions.[69] The aforementioned blind spots in the book attest to the wide-ranging breadth and complexity of the processes that drive resource extraction in the contemporary world. In my view, however, they can also be seen as invitations and open questions for future research under the conceptual frameworks hereby proposed.

Chapter 2 begins by proposing a reading of resource imperialism that departs from the statism and Eurocentrism of traditional approaches, and whose level of analysis is not "the West" or the international political relations of the nation-state but the world market as a totalizing sociomaterial system. With this, I reclaim a scholarly tradition that has considered class relations to precede the existence of the state in the configuration of the capitalist world-system. An approach that makes analytical distinctions between the essential content of global capitalism (the self-expansion of value through the material transformation of the labor process) and its multiple historical and phenomenal appearances in the *form* of institutional politics (such as militarization, debt peonage, interventionism, etc.), is adequately positioned to interpret the nature of imperialist/subimperialist practice in the current postcolonial context. The chapter begins with a brief exploration of the scientific-technological revolutions that have enabled Western imperial powers to secure

---

Marginalisation of Post-Extractivist Forces in Evo Morales's Bolivia," *Geoforum* 84 (2015): 280–91; Jeffery R. Webber, "The Indigenous Community as 'Living Organism': José Carlos Mariátegui, Romantic Marxism, and Extractive Capitalism in the Andes," *Theory and Society* 44, no. 6 (2015): 575–98; Japhy Wilson and Manuel Bayón, "Concrete Jungle: Planetary Urbanization in the Ecuadorian Amazon," *Human Geography* 8, no. 3 (2015): 1–23; Thomas Purcell, "The Political Economy of Rentier Capitalism and the Limits to Agrarian Transformation in Venezuela," *Journal of Agrarian Change* 17 (2017): 296–312.

69   According to Eduardo Gudynas, although the post-neoliberal governments of Latin America's Pink Tide have enacted frameworks for natural resource governance that reject some elements of neoclassical economics, they share the predilection for "growth," "progress," and "development" that is typical of neoliberal governments. In general terms, the model of economic growth that underpins these governments seeks to ensure consumption and the expansion of mainstream economic indicators such as gross domestic product. Eduardo Gudynas, "Value, Growth, Development: South American Lessons for a New Ecopolitics," *Capitalism Nature Socialism*, September 11, 2017, tandfonline.com/doi/abs/10.1080/10455752.2017.1372502.

access to raw materials across historical cycles of accumulation. I then bring the proposed theoretical approach to life by exploring the spaces of extraction that have emerged as the Pacific Ocean becomes the pivot of world trade, and the "Asian Tigers" assert themselves as the world's main buyers of raw materials and main creditor nations.

Chapter 3 continues to interrogate the present condition of resource extraction through an emphasis on the geographies of labor resulting from an upgraded technological basis of mineral production and circulation. Through the specific case of the mining supply chain in Chile, and from a value-theoretical perspective, this chapter challenges the idea that automation will spell the end of work. It shows how the expansion of the productive attributes of the workers in charge of the most complex parts of the mining process (geological modeling, engineering, programming of equipment, etc.) is directly connected to the degradation of those of an ever-increasing contingent of precarious, non-wage, racialized, and gendered workers. The productive whole that results from the internal polarization of the productive capacities of the organs of the collective laborer, I suggest, is not consciously regulated but becomes a form of existence of capital. With this, I reveal the centrality of the commodification of labor-power to the capitalist production of space. The exploitation of a diverse multiplicity of concrete labors manifests itself spatially through the production of an uneven, expanding fabric of urbanization: cosmopolitan, green, and smart cities for the highly skilled workers of the mining industry; polluted towns, shanty towns, and overcrowded camp sites for its invisible workforces and its expanding surplus populations.

In chapter 4, I interrogate the relevance of the logistics revolution to the organizational restructuring of resource-based industries, especially as the Pacific Ocean becomes one of the main infrastructural corridors for the circulation of raw materials. Through the case of port cities in northern Chile, this chapter suggests that the logistics turn in the extractive industries heralds a more advanced degree of functional integration between resource extraction, port operations, and maritime transport, rendering a complex infrastructural system that connects advanced manufacturing in Asia with large-scale mining in Latin America. Through a materialist reading of the capitalist state, the paper suggests that the metabolic flows that animate this complex socioecological system are mediated by distinct, yet overlapping tendencies toward the internationalization and the concentration of the political authority of the late liberal state. Moreover, the seamless movement of the logistical systems of extraction across national borders has been cemented on what I term *logistical urbanization*: a new rubric of macrospatial infrastructural planning and territorial design that revolves around speed, continuity, and connectivity, and where city and noncity space are cast in relational terms.

Through an analysis of Chile's neoliberal technocracy, chapter 5 continues to interrogate the role of the state in the production, expansion, and reproduction of primary-commodity frontiers. This is carried out by presenting an alternative reading of systematic primitive accumulation that is sensitive to law-making violence and forceful expropriation but ultimately posits them as *subservient* to and *ontologically distinct* from the impersonal forms of social mediation intrinsic to modern society. Chapter 6 explores the messy entanglements between global monetary space, transnational mining companies, and everyday environments in geographies of extraction. The chapter suggests that the liquidity required to expand the scale of mineral production through advanced mechanization, vertical reintegration, and organizational restructuring, has been garnered by intricate debt instruments and financial arrangements operating at various spatial scales—from the issuance of sovereign bonds and derivatives contracts, to the financialization of the household. Moreover, the chapter also reveals how the financialization of the mining industry has inadvertently extended the discipline of monetary relations to the mining towns of northern Chile, transforming previously self-subsistent households and peasants into wage-laborers and debtors. As capital's archetypal form of impersonal domination, money spearheads the irruption of value relations into the countryside, so the expansion of spatial technologies of primary-commodity production in Chile has evolved alongside the dissolution of rural and agrarian ways of life by means of the cash-nexus.

Chapter 7 reflects on the relationship between technological modernization and the revolutionary subjectivity of the working class. I build upon the late Marx's partially published writings on precapitalist and non-Western societies as a means to understand the radical sociocultural heterogeneity that underlies the global collective laborer in the twenty-first century. In these little-known writings, Marx steers away from the implicit superiority he had accorded to European societies and begins to identify, in the "actually existing communisms" of non-Western/premodern societies, some of the latent and yet unrealized elements of a future postcapitalist civilization. These communal forms, according to Marx, had been constricted to a self-contained and parochial existence, so he considered modern science and technology the vital force that could scale up such ancient modes of social metabolism into an advanced and universal civilizational form. Against the backdrop of these texts, the chapter then illustrates how the technological infrastructures of extraction have allowed subaltern struggles—of subcontracted workers and indigenous, peasant, and women's movements—to supersede their isolation and project their communal social rationality across scales. Most importantly, this chapter highlights a very similar process taking place throughout the manufacturing centers of Asia, themselves connected by myriad metabolic flows to sites of extraction.

In the concluding chapter, I assess the core contributions of the book and reflect on the potential of the present industrial era to bring about the ferment for a postcapitalist scientific praxis. I tease out the idea that the Marxian vision of communism contains a pastoral design whereby the lost unity between algorithmic knowledge and folk knowledge (and thus between science and life) is set to be restored within the context of an advanced technological future. As global supply chains lead a race to the bottom in terms of labor standards, further eroding the boundaries between city and noncity and intellectual and manual laborers, the material conditions for such utopian vision become ever more feasible. To the extent that scientific labor-power becomes aggressively proletarianized and the fragmentation of the sciences makes "expert idiocy" the hallmark product of the neoliberal university, some intellectual laborers have rebelled, asserting their technical expertise as a medium for the conscious organization of transformative action.[70] The logistical space of extraction has then emerged as a melting pot for new alliances between scientists and peasants, engineers and workers, lawyers and

---

70 An "expert idiot" or *idiot savant* is an individual who is proficient at mobilizing highly specialized algorithmic knowledge yet unable to understand its imbrication within the broader dynamics of economy, polity, and society. Mandel, *Late Capitalism*, 261; Dimitris Milonakis and Ben Fine, *From Political Economy to Economics: Method, the Social and the Historical in the Evolution of Economic Theory* (New York: Routledge, 2009), 12; Ursula Huws, *Labor in the Global Digital Economy: The Cybertariat Comes of Age* (New York: Monthly Review Press, 2014).

indigenous leaders, artists and precarious migrants. Although fragmentary and still in their infancy, such budding dialogues and cross-fertilizations are drawing the blueprints for a new type of scientific consciousness, one that will enable humans and machines to supersede their inverted existence as character masks of blind and unruly powers and work collectively for the conscious, genuinely democratic regulation of their environments.

2. EMPIRE

Resource Imperialism after the West

It may be asked whether the theory of imperialism should not take the world market as the *a priori* level of analysis from which conclusions might be drawn, rather than taking national capital and the state associated with it as its starting point.[1]

Something larger than evil rules over these worlds.[2]

Power is the very organization of this world, this engineered, configured, *purposed* world. That is the secret, *and it's that there isn't one.*[3]

Introduction

Capitalism is unique, Neil Smith considered, in that for the first time in history, human beings produce nature at a world scale.[4] When one looks at the sheer technological sophistication and the magnitude at which the mining industry wrests minerals from the soil to swiftly move them

---

1   Braunmühl, "On the Analysis of the Bourgeois Nation State," 162.

2   William Ospina, *Poesía* (Bogotá: La Otra Orilla, 2008), 266.

3   Invisible Committee, "To Our Friends," *Semiotext(e)*, 2015, 83. Emphasis in original.

4   Neil Smith, *Uneven Development: Nature, Capital and the Production of Space* (Athens: University of Georgia Press, 2008 [1984]), 77.

around by air, land, and sea, it becomes possible to start grasping the full extent of Smith's provocative assertion. The relentless robotization and computerization advanced by the mining industry during recent decades makes almost any other sector of social production today seem rudimentary at best. Although Google engineers have been testing prototypes for a self-driving car that could tentatively be released into the market at some point during the 2020s, mining companies have been operating with fully robotized vehicles since 2008. These driverless trucks, pioneered by BHP in association with the Japanese giant Komatsu, are fully autonomous and dwarf, in size and cargo capacity, any type of wheeled haulage machinery.[5] Besides autonomous trucks, mining companies have harnessed advances in robotics, control systems, and materials science in order to mechanize and computerize parts of the extraction process. This has allowed them to introduce automatic drills, smelters, locomotives, cranes, and other technological elements to diverse segments of the supply chain. Moreover, the introduction of geospatial information systems (GIS), artificial intelligence, and geological modeling tools to mineral forecasting has allowed companies to extract low-grade ore bodies profitably for the first time in history, especially without the burden, timescales, and costs of drilling boreholes.[6] By making use of GIS, electromagnetic waves, and 3-D visualization methods imported from videogaming technologies, geologists and engineers can now develop very accurate representations of the subsurface in order to design the most effective mine plan.[7]

According to Ernest Mandel, each epochal shift in capitalist society demands a qualitative leap forward in the technical process, which can only be attained by means of a new generation of machines.[8] Major

---

5   Kathryn Diss, "Robotic Trucks Taking Over Pilbara Mining Operations in Shift to Automation," ABC Australian Broadcasting Corporation, April 25, 2014, abc.net.au; Minería Chilena, "Camiones autónomos en Gaby: Productividad aumentó en 25%," January 28, 2014, mch.cl/reportajes/camiones-autonomos-en-gaby-productividad-aumento-en-25.

6   Interview with a geophysicist from London Mining Network, September 11, 2015; Interview with a geologist from Lhoist, March 8, 2016.

7   Manaugh, "Infinite Exchange"; Interview with a mining company engineer, March 9, 2016.

8   Mandel, *Late Capitalism*, 110. For Mandel, the fundamental revolutions in the technology of the production of motive machines by machines (i.e., machinofacture) constitute the determinant moment in technological revolutions as a whole. Accordingly, he outlines the following three technological revolutions or developmental landmarks

theories of global political economy in the Marxist tradition have typically considered epoch-making shifts and technological revolutions of the type Mandel described to go hand in hand with a new structure of geopolitical relations. Such relations are typically understood as driven by empire-building projects whereby a new "hegemon" mobilizes the powers of science and technology in order to achieve trade dominance. This was the claim advanced by the influential accounts of world-systems analysis[9] as well as by the related approaches of dependency theory and Latin American structuralism.[10] Studies of capitalism in the *longue durée* have associated the existing resource-extraction frontiers—sugarcane in the sixteenth century, peat in the eighteenth, rubber and coal in the nineteenth, oil and iron ore in the twentieth—with the pursuit for world dominance by Western imperial powers. The so-called "fourth industrial revolution,"[11] considered by pundits to be an era of technological innovation whose breadth and dynamism supersede those of previous historical epochs, seems to be lacking its proverbial hegemon. Yet, paradoxically, this allegedly postcolonial, postpolitical context has witnessed the expansion of mineral-extraction frontiers and the concomitant clearing of peasantries from the land to an extent that is entirely without precedent in human history.

This chapter sets out to solve such a paradox by arguing that existing studies have tended to confuse the political/historical forms of appearance of capitalist imperialism with their underlying content in the production and valorization of value, a process whose existence not only transcends the political mediation of domestic spheres of accumulation but *is ontologically prior* to them. The purpose of the chapter is

---

in the history of large-scale industry: machine production of steam-driven motors since 1848 (external combustion); machine production of electric and (internal) combustion motors since the late nineteenth century; and machine production of electronic and nuclear-powered apparatuses from the 1940s on.

9   Immanuel Wallerstein, *The Politics of the World Economy: The States, the Movements and the Civilizations* (Cambridge: Cambridge University Press, 1991 [1984]); Arrighi, *Long Twentieth Century*; Bunker and Ciccantell, *Globalization*.

10   For an overview, see Cristóbal Kay, "Teorías latinoamericanas del desarrollo," *Nueva Sociedad* 113 (1991): 101–13; Cristóbal Kay, "Estructuralismo y teoría de la dependencia en el período neoliberal," *Nueva Sociedad* 158, 1998, 100–119.

11   Bloem et al., *Fourth Industrial Revolution*; Brynjolfsson and McAfee, *Second Machine Age*; Maynard, "Navigating the Fourth Industrial Revolution"; Schwab, *Fourth Industrial Revolution*.

therefore to posit the world market (not the nation-state or even the interstate system) as the analytical starting point from which the nature of resource imperialism can be most adequately fleshed out. This entails an analytical dissection of the "fetishized" or "alienated" imperialist political forms, which are sensuous and fragmented (e.g., militarization, debt peonage, internal colonialism, dependency), from their essential foundations in the movement of value, a process that is suprasensuous and systemic. Philosophically speaking, this entails capturing how the essential level (the total surplus value of society) acquires phenomenal reality in sensuous experience via the messy materialities and entanglements of firms and states. The reading proposed here is thus inspired by Marx's appropriation of the Hegelian conception of the inverted world, which posits reality as the unity of two contradictory movements.[12] For Marx, capital is a "sensuous supersensible thing." This means that the reality of liberal society is a product of the movement of opposites, between self-determined activity and its independent appearance in the autonomized forms of political power.[13] The categorical critique that this chapter proposes involves deciphering the practical and human content that underlies such alienated forms.

To develop a reading of the production of resource frontiers in the context of global capital accumulation, I build upon value-theoretical

---

12 Hegel's monist metaphysics constitutes a fundamental departure from the traditional "two-world model" of European philosophy, which goes back to Plato's famous allegory of the cave. For an overview of Hegel's rubric of monist metaphysics, see Frederick C. Beiser, "Introduction: Hegel and the Problem of Metaphysics," in Frederick C. Beiser (ed.), *The Cambridge Companion to Hegel* (Cambridge: Cambridge University Press, 1993; Robert Pippin, "You Can't Get There from Here: Transition Problems in Hegel's Phenomenology of Spirit," in Beiser (ed.), *Cambridge Companion to Hegel*. Hegel's naturalistic understanding of the absolute as a living organism of sorts, allowed him to supersede dualistic formulations inherited from Western philosophy, which conceived of mind and body, nature and society, etc., as different substances. In Hegel's idea of a single world, by contrast, subject and object are not distinct entities, but different degrees of organization of the same living force (Beiser, "Introduction"). The revolutionary novelty of Hegel's philosophy, according to Helmut Reichelt, then stems from the fact that he considered reality as appearance; as a single, inverted world that contains both essence and existence. Helmut Reichelt, "Social Reality as Appearance: Some Notes on Marx's Conception of Reality," in Werner Bonefeld and Kosmas Psychopedis (eds.), *Human Dignity: Social Autonomy and the Critique of Capitalism* (Aldershot, UK: Ashgate, 2005).

13 Postone, "Time, Labor, and Social Domination" and "Lukács and the Dialectical Critique of Capitalism"; Reichelt, "Social Reality as Appearance."

interpretations of the world economy whose methodological approach has consisted of a logical progression from the determination of the total surplus value of society—the world market—to its organization into individual parts—national economies and individual capitals.[14] Some of these approaches, the chapter shows, have emerged not only from the form-analysis tradition, but also from a radical strand of Latin American theories of dependency, which has considered class relations to precede those of the nation-state.[15] On this basis, and as opposed to dominant readings, the chapter argues that resource imperialism is not autonomously determined by the locational strategies of transnational firms or by the political dynamics of the nation-state. According to Vivek Chibber, one of the most salient aspects of the classical theories of imperialism that emerged in the context of the Second International was that they sought to decipher the deep economic forces that under-pinned what on the surface appeared to be autonomous political projects.[16] With this, I intend to shift the focus from political theories of imperialism to those that place a greater emphasis on economic and systemic determinations. Accordingly, developing a theory of imperial-ism that takes seriously the essential unity of global capital accumula-tion is a matter of intellectual and political urgency, especially in the context of a new international division of labor that destabilizes the geometries of power of an interstate system originally structured around global North/global South and West/non-West binaries.[17]

In the first section of the chapter, I briefly review how major intellectual traditions have traditionally considered the making of resource peripheries as linked to empire-building projects and,

---

14 Braunmühl 1978, "On the Analysis of the Bourgeois Nation State"; Dussel, *Towards an Unknown Marx*; Iñigo Carrera, *El Capital*; McNally, "Blood of the Commonwealth"; Fred Moseley, *Money and Totality: A Macro-Monetary Interpretation of Marx's Logic in Capital and the End of the "Transformation Problem"* (Chicago: Haymarket, 2017); Gastón Caligaris, "Los países productores de materias primas en la unidad mundial de la acumulación de capital: un enfoque alternativo," *Cuadernos de Economía Crítica* 6 (2017): 15–43.

15 Dussel, *Towards an Unknown Marx*; Ruy Mauro Marini, *Dialéctica de la dependencia* (Mexico: Era, 1973); Agustín Cueva, *Teoría social y procesos políticos en América Latina* (Mexico: Edicol, 1979).

16 Chibber, "Return of Imperialism."

17 Fröbel et al., *New International Division of Labour*; Mezzadra and Neilson, *Border as Method*; Charnock and Starosta, *New International Division of Labour*; Starosta, "Revisiting the New International Division of Labour Thesis."

more generally, to the direct political-economic relations of an interstate system. By means of Mandel's periodization of industrial capitalism,[18] I excavate the scientific-technological revolutions that have enabled access to raw materials across previous historical cycles of accumulation. The second section goes on to assess the historical specificity of what I term the *fourth machine age*. This advance in modern science and technology, I argue, has been crucial in the processes that have repositioned the gravitational center of the world economy toward the Pacific Ocean. In the third section, the chapter builds upon value-theoretical readings of global capitalism in order to lay out an alternative framework of resource imperialism that can grasp the nature of capitalism as a planetary socionatural system but also takes seriously the evolving forms of political authority and extraeconomic force that mediate its complex metabolism. The final section grounds and spatializes these theoretical insights by exploring the spaces of extraction that have emerged as the Asian Tigers consolidate themselves as the world's main buyers of raw materials.

Empire and Technologies of Extraction

An exploration of the colonial histories and geographies of the last six centuries reveals how natural-resource frontiers are internally related to the constitution of the very fabric of modernity. Without the fabulous material wealth drawn from the sugarcane plantations of Brazil and the silver mines of Potosí (now in Bolivia) in the sixteenth century, for example, the cultural, artistic and political efflorescence that characterized the so-called Golden Century of the Habsburg dynasty in the Iberian Peninsula would have never existed. Likewise, the first industrial revolution that took place in nineteenth-century England would have been unthinkable without the rubber, guano, and coal frontiers that dramatically expanded across the Atlantic Ocean in order to feed machines, crops, and workers in the heartland of the British Empire. World-systems analysis is perhaps one of the most, if not the most, influential intellectual traditions explaining such relations of interdependence in the

---

18   Mandel, *Late Capitalism*.

configuration of the space economy of capitalism. Immanuel Wallerstein, a key proponent of this strand of thought, starts from the assumption that states are the expression of power in a capitalist world economy as they enforce the appropriation of value from the bourgeois class.[19] As a fractured and uneven system, such appropriation of value unfolds along constant pressure from the strong against the weak, and thus a polarized system of "core" and "peripheral" states is summoned into existence.[20] The political relations of imperialist expansion, so the argument goes, translate into economic relations of unequal exchange between cores and peripheries.[21]

Such world systems are dispersed across space but also across time, and for this reason one of the key features of world-systems analysis is its opposition to the so-called "two-century model" that views the capital form as an offspring of the first industrial revolution of the nineteenth century. Giovanni Arrighi's influential account of systemic cycles of accumulation explains the genesis and evolution of world systems in the *longue durée*, with the fifteenth century as its starting point. For him, the initial formation and subsequent expansion of the world system to its present global all-encompassing dimensions can be broken down into four, partly overlapping systemic cycles of accumulation: a Genoese-Iberian cycle that stretches from the fifteenth through the early seventeenth; a Dutch cycle, stretching from the late sixteenth century to the late eighteenth; a British cycle, stretching from the mid-eighteenth century to the early twentieth; and a US cycle, stretching from the late nineteenth century to what he saw as the wave of economic expansion taking place in the late twentieth and early twenty-first.[22] In Arrighi's formulation, a systemic cycle is superseded once an emergent core state is able to consolidate itself by means of material and financial expansion and achieve trade dominance.

The inherited epistemological frameworks and historical assumptions introduced by world-systems analysis have been fundamental to how primary-commodity production has been understood across disciplines and intellectual traditions. Stephen Bunker and Paul Ciccantell's

19 Wallerstein, *Politics of the World Economy*.
20 Ibid., 5.
21 Ibid.
22 Arrighi, *Long Twentieth Century*.

"new historical materialist" study of natural resource frontiers, for example, is directly anchored to Arrighi's formulation of systemic cycles of accumulation.[23] However, Bunker and Ciccantell depart from Arrighi's reading because they place the gravitational focus not on finance but on primary commodities. The crux of the question, these authors argue, lies in the capacity of ascending imperial powers to secure and maintain access to raw materials through scientific and technological innovation.[24] As industries in the "core" become more capitalized and the ratio of dead labor to living labor rises along with productivity, access to an increasing amount of resources in increasingly remote "peripheries" needs to be secured. This imposes the need to reduce transport costs, so a characteristic feature of each systemic cycle of accumulation is that the ascending economy is able to introduce technological innovations that allow for an increase in the scale and efficiency of transport.[25] Larger and more efficient ships, ports, railways, warehouses, and other forms of transport infrastructure, according to Bunker and Ciccantell, have played fundamental roles in the competition of states for global trade dominance.

The ascent of the Dutch to trade dominance, for example, was to a large extent a result of the introduction of technologies to maneuver oak and pine wood in order to build lighter and more efficient hulls—the Dutch *fluyt*.[26] In this dawn of modern technics, which Lewis Mumford terms the *eotechnic phase*, the water-and-wood industrial complex set the foundations for experimental science on mathematics, exact measurement, and timing.[27] The shift to the *paleotechnic phase*—which in Mandel's periodization corresponds to the first technological revolution—built upon the previous scientific revolutions and inaugurated a coal-and-iron complex that relied on new resources such as aluminum, cassiterite, manganese, petroleum, and rubber. The production of automatic systems of machinery feeding upon and at the same time

23   Stephen Bunker and Paul Ciccantell, "The Economic Ascent of China and the Potential for Restructuring the Capitalist World-Economy," *Journal of World-Systems Research* 10, no. 3 (2004): 565–89; Bunker and Ciccantell, *Globalization*.

24   Bunker and Ciccantell, *Globalization*.

25   Ibid.

26   Ibid., 10.

27   Lewis Mumford, *Technics and Civilization* (Chicago: University of Chicago Press, 2010 [1934]); Alfred Sohn-Rethel, *Intellectual and Manual Labor: A Critique of Epistemology* (London: Macmillan, 1978); Moore, *Capitalism in the Web of Life*.

expanding these new resource frontiers, Bunker and Ciccantell argue, allowed the British to achieve trade dominance and set into motion a new systemic cycle of accumulation.[28] Innovations in motor design allowed them to introduce mechanized ships that gradually but irrevocably doomed the sailing ships inherited from the previous systemic cycle.[29] This opened new possibilities for expanding resource peripheries in more remote geographies. Rubber, in particular, performed key mechanical functions in conveyor belts, pads for moving parts that rubbed against each other, insulation for cables, and tires for wheels that made machines mobile. The rubber boom that followed the consolidation of the British Empire, Bunker and Ciccantell note, vastly reconfigured the geography of the Amazon, producing major social and environmental destruction.[30] The British Empire remained unchallenged until the United States pioneered the process of Bessemer conversion for iron-ore smelting, which made durable steel that was cheap enough for mass production.[31]

Bessemer steel production facilitated unprecedented mechanization of agriculture, extraction, and industry, as well as the rapid transport of raw materials that consolidated the US as a new imperial power after the mid-nineteenth century.[32] Maximum ore cargos increased from 1,000 tons in the 1870s to 3,000 tons when the first steel ships were built in 1886.[33] The invention and proliferation of the internal-combustion engine, which for Mandel marks the "second technological revolution," allowed the US to further improve transport technologies and substantially reduce the turnover times of capital. Iron-ore mines swelled in size and grew in numbers after these key technological breakthroughs, which ensured US hegemony until Japan devised new computerized technologies for iron-ore smelting, which in turn dramatically improved ships in both propulsion and cargo capacity. The systemic cycle that Bunker and Ciccantell ascribe to the ascendancy of Japan corresponds to the "third technological revolution" (electronic and nuclear-powered apparatuses) in Mandel's periodization.

28   Bunker and Ciccantell, *Globalization*.
29   Ibid., 160.
30   Ibid., 38.
31   Ibid., 40.
32   Ibid., 40–41.
33   Ibid., 170.

Despite their differences in scope and method, what cuts across these world-systems perspectives on resource extraction is that they are underpinned by a deeply rooted methodological nationalism that views these historical transformations as a result of interactions between states or systems of states. In general terms, methodological nationalism has been understood as a metatheoretical orientation that conflates society with the state and the national territory. Most importantly, though, it has also been understood as an approach that isolates internal and external factors in explanations of development, giving more prevalence to the former.[34] As Brenner suggests, although Wallerstein's concept of the modern world system is framed on the basis of an attempt to supersede state-centric models of capitalist modernity, national territories remain pivotal within the whole theoretical edifice.[35] Although the division of labor in the capitalist world economy is considered to be structured in accordance with three supranational zones (core, periphery, semiperiphery), Wallerstein's reading consistently places the focus on the specific historical trajectories and dynamics of nation-states. Transnational corporations, infrastructural megadevelopments, and circuits of capital, according to Brenner, remain secondary in Wallerstein's approach.[36] In the end, the primary geographical units of global space remain defined by the territorial boundaries of domestic spheres of accumulation.

A very similar methodological and metatheoretical orientation informs other predominant approaches to natural-resource governance and extraction, such as theories of natural-resource curse, ecological economics, and the Latin American schools of structuralism, dependency theory, and *post-extractivismo*. Latin American structuralism is perhaps the most influential intellectual tradition of this group. Its most renowned author was the Argentinean economist Raúl Prebisch, who also served as the executive director of the United Nations' Economic Commission for Latin America and the Caribbean (ECLAC) in the

---

34   Daniel Chernilo, "The Critique of Methodological Nationalism: Theory and History," *Thesis Eleven* 106, no. 1 (2011): 98–117; Lucia Pradella, "New Developmentalism and the Origins of Methodological Nationalism," *Competition and Change* 18, no. 2 (2014): 180–93.

35   Neil Brenner, *New State Spaces: Urban Governance and the Rescaling of Statehood* (Oxford: Oxford University Press, 2004).

36   Brenner, *New State Spaces*, 51.

1950s. Prebisch's ideas, it should be pointed out, were formative in Wallerstein's theorization of the modern world system[37] and in the theories of dependency that emerged from the 1960s onward.[38] The basic tenet of the structuralist framework, according to Cristóbal Kay is that international commerce reproduces the inequalities between the center and the periphery.[39] This theoretical framework was devised by ECLAC structuralist economists to make sense of the incorporation of Latin American nations into the international division of labor as suppliers of raw materials. The common thread that cuts across all such approaches, according to recent critiques, is a presupposition of the nation-state as internally constituted by its own domestic context.[40] Accordingly, international commerce is therefore construed as being the process of interaction between these abstractly autonomous spheres of national accumulation. As Enrique Dussel explains, Latin American theories of dependency gave prevalence to the surface appearance of dependency— i.e., its historical manifestation in underdeveloped economies. This, in his view, led to a state-centric reading of the interstate system that was oblivious to the operation of the law of value on a world scale—its essence.[41]

---

37 On the influence of Latin American structuralism on world-systems analysis, see Aníbal Quijano, "Coloniality of Power, Eurocentrism, and Latin America," *Nepantla* 1, no. 3 (2000): 533–80; Walter Mignolo, "The Geopolitics of Knowledge and the Colonial Difference," in Mabel Moraña, Enrique Dussel, and Carlos Jáuregui (eds.), *Coloniality at Large: Latin America and the Postcolonial Debate* (Durham, NC: Duke University Press, 2008).

38 Kay, "Estructuralismo"; Dussel, *Towards an Unknown Marx*, chapter 13.

39 Kay, "Teorías latinoamericanas."

40 Juan Grigera, "Conspicuous Silences: State and Class in Structuralist and Neostructuralist Thought," in Susan Spronk and Jeffery Webber (eds.), *Crisis and Contradiction: Marxist Perspectives on Latin America in the Global Political Economy* (Leiden: Brill, 2014); Caligaris, "Los países productores." It is very important to point out that there are substantial internal nuances between Latin American theories of dependency. Although the "reformist" reading of dependency proposed by Raúl Prebisch, Celso Furtado, and Aníbal Pinto at ECLAC became the most influential, there was also a "revolutionary" reading of dependency, premised on a more nuanced reading of the state. The main proponents of the "revolutionary" approach to dependency are Ruy Mauro Marini, Agustín Cueva, Vania Bambirra, and Enrique Dussel.

41 Dussel, *Towards an Unknown Marx*.

The Fourth Machine Age and the Rise of
an Asia-Centered World System

The rise of China and other Asian Tigers, coupled with the demise of formal colonization after the 1970s, puts into question the types of state-centric and core–periphery readings of the capitalist world system that have informed major studies of extraction. One of the most striking particularities of the commodity supercycle of recent decades is that, for the first time in modern history, the vast material wealth that is wrested from mines, oil wells, and croplands, is shipped to countries tradition-ally considered "peripheral" or "semiperipheral." The stunning economic growth and industrial transformation of the various generations of East Asian economies, combined with the emergence of the BRICS countries (Brazil, Russia, India, China, South Africa)—which command a grow-ing share of world trade—has expanded the volume of raw materials circulating in both financial and spot markets and shifted its geographi-cal focus toward a more "South-South" configuration.[42] The exponential increase in manufactured goods traded at the global level, which rose from around $2 trillion in 1980 to nearly $16 trillion in 2000,[43] attests to this cyclopean shift in the scale of capitalist production. The geographi-cal distribution of this increase in manufacturing capacity is why some commentators now refer to the twenty-first century as the "Pacific Century."[44] From 1990 to 2012, Asia's share of global manufacturing rose from 25 percent to 50 percent, and it is estimated that this figure will continue to grow during the next decades.[45]

East Asian economies, initially considered mere export-processing zones for Western transnational corporations,[46] have managed to revo-lutionize instruments and relations of production and emancipate

---

42   Pádraig Carmody, *The New Scramble for Africa* (Cambridge: Polity, 2011).

43   Nikos Katsikis, *From Hinterland to Hinterglobe: Urbanization as Geographical Organization*. Unpublished PhD thesis submitted to Harvard University, 2016, 302. All monetary amounts are in US dollars unless otherwise indicated.

44   Thomas Wilkins, "The New 'Pacific Century' and the Rise of China: An International Relations Perspective," *Australian Journal of International Affairs* 64, no. 4 (2010): 381–405.

45   Khanna, *Connectography*, 89.

46   Fröbel et al., *New International Division of Labour*; Mies, *Patriarchy and Accumulation*.

themselves from "captive" global supply chains.[47] China, a relative late-
comer to this process, has also been successfully combining export-
oriented industrialization with an emphasis on "indigenous innovation"
(*zìzhǔ chuàngxīn*).[48] This, it has been argued, has altered the power
dynamics and the governance composition of global manufacturing in
ways that no one would have foreseen even a few decades ago.[49] In other
words, China and other East Asian economies are no longer mere recipients of foreign direct investment advanced for the development of
special economic zones or world-market factories specialized in
low-value-added, unskilled labor (a "semiperiphery" in Wallerstein's
phraseology). Rather, they have been at the forefront of technological
innovation and industrial upgrading based on the aggressive capitalization and robotization of the production line. China's "robot revolution,"
which consists of introducing fully autonomous factories that operate
without need of lights (as opposed to humans, robots can work in the
dark), is one of the latest installments in the modernizing project

----

47   It was such industrial upgrading that made Fröbel, Heinrichs, and Kreye's thesis
of the "new international division of labor" fall out of favor in the 1990s. For these
authors, industrialization in the global South was limited to unskilled, labor-intensive
productive processes, and hence the new geography of industrialization did not
overcome old relations of dependency between these newly industrializing countries
and the Western core.

48   One of the signature fields of specialization in Chinese innovation has been
nanotechnologies, and specifically "microelectromechanical systems" (MEMS).
Although MEMS technologies are still in their infancy, they are already being successfully
implemented in a wide array of industrial applications and consumer products, such as
automobile airbags, smartphones, control systems for agriculture and primary-
commodity production, and the health and transport sector. The novelty of these
nanotechnologies is that they help reduce consumption of thermal and electrical energy
in mechanical and electronic systems. With nearly 3,500 researchers and 200 research
centers, the Chinese government has embraced MEMS technologies in recent years as
one of the country's flagship fields of scientific-technological innovation. Yalú Maricela
Morales Martínez, "La política actual de ciencia y tecnología de China en la tecnología
MEMS o sistemas micro-electromecánicos," in Dussel Peters (ed.), *América Latina y el
Caribe-China*.

49   Richard Appelbaum, "Big Suppliers in Greater China: A Growing Counterweight
to the Power of Giant Retailers," in Hung (ed.), *China and the Transformation of Global
Capitalism*; Alvin So, "Rethinking the Chinese Developmental Miracle," in Hung (ed.),
China and the Transformation of Global Capitalism; Morales Martínez, "La política
actual de ciencia y tecnología de China en la tecnología MEMS o sistemas micro-
electromecánicos," in Dussel Peters (ed.), *América Latina y el Caribe-China: Economía,
Comercio e Inversiones*.

inaugurated by the Eleventh Five-Year Plan (2006–2011) and in the Long-Term Plan for the Development of Science and Technology.[50]

The rise of China, according to Ho-fung Hung, marks a clear turning point in geopolitical configurations inherited from classical international divisions of labor because, unlike earlier "tigers" (such as Japan, Taiwan, and South Korea), China is not a client or vassal state of the US but has emerged as "an independent geopolitical and military force capable of challenging the United States."[51] Besides military prowess, Chinese financial institutions now rank among the most powerful in the world, and China's monetary system has become a strong counterweight to international financial institutions involved in sovereign lending, such as the World Bank, the International Monetary Fund, and the Inter-American Development Bank.[52] Underlying this major transfer of geopolitical power from the West to the non-West is the leap forward in the automation and computerization of the labor process that Klaus Schwab, founder of the World Economic Forum, terms "the fourth industrial revolution."[53] Given its scale, scope, and complexity, Schwab asserts that the present industrial revolution is "unlike anything humankind has experienced before."[54] Its distinctiveness does not necessarily emerge from novel technological innovations such as mobile electronic devices, digital fabrication, artificial intelligence (AI), big data, nanotechnology, biotechnology, the internet of things, robotics, materials science, and quantum computing. Although still in their infancy, such technologies have already made key breakthroughs across economic, cultural, and social realms. For Schwab, the specificity of the fourth

50   Appelbaum, "Big Suppliers in Greater China: A Growing Counterweight to the Power of Giant Retailers," in Hung (ed.), *China and the Transformation of Global Capitalism*; Will Knight, "China Is Building a Robot Army of Model Workers," *MIT Technology Review*, April 26, 2016, technologyreview.com; Ben Bland, "China's Robot Revolution," *Financial Times*, April 28, 2016, ft.com.

51   Hung, *China and the Transformation of Global Capitalism*, 13.

52   Leonardo Stanley, "El proceso de internacionalización del RMB y el nuevo protagonismo del sistema financiero chino," in Dussel Peters (ed.), *América Latina y el Caribe-China*; Eric Helleiner and Jonathan Kirshner, "The Politics of China's International Monetary Relations," in Eric Helleiner and Jonathan Kirshner (eds.), *The Great Wall of Money: Power and Politics in China's International Monetary Relations* (Ithaca, NY: Cornell University Press, 2014).

53   Schwab, *Fourth Industrial Revolution*; Brynjolfsson and McAffee, *Second Machine Age*.

54   Schwab, *Fourth Industrial Revolution*, 1.

industrial revolution stems from the systematic fusion and interaction of such technologies across the physical, digital, and biological domains.

In the mining industry, the productive articulation between robotics, biotechnology, AI, GIS, and control systems has revolutionized the process of extraction as we know it. For example, an autonomous mining truck operates with an average of 1,000 sensors. This allows it to maintain constant communication with its environment, with other machines, and with engineers working across other phases of the extraction process.[55] Another example is that the employment of engineered microorganisms to break the resistance of recalcitrant ores that cannot be extracted through traditional methods, as Labban shows, has rendered mining a biologically based industry and has even extended the process of extraction at the cellular-elemental scale.[56] Yet, for synthetic bacteria to be applied to the extraction of low-grade ore bodies, a large-scale haulage system with the capacity to mobilize massive volumes of rock first needs to be in place. For autonomous trucks and shovels to even begin to move large volumes of rock, in turn, mineral deposits must first be made "legible" by GIS and sophisticated geological-modeling and data-processing tools. In other words, the capital-intensive mine is not an exclusive product of robots, GIS, microorganisms, control systems, or laborers, but of the synergistic integration between the productive capacities of these human and extrahuman elements. As a result, mineral deposits that had not been mined because they were "uneconomical" under older technologies are now being reopened and transformed into large mines across every corner of the world, putting enormous pressure upon water sources, livelihoods, and communities.[57]

The ability to mine low-grade mineral deposits has made the mining industry more profitable and has increased the material footprint of mineral extraction by a factor of around 1,000 (in terms of the ratio of solid waste produced per gram of mineral extracted).[58] To the extent that the increase in labor productivity enabled by these technologies has

55   Simonite, "Mining 24 Hours a Day with Robots"; interview with a Chilean mining executive, January 10, 2017.

56   Labban, "Deterritorializing Extraction."

57   Swift, "Stop the Gold Rush!"

58   Interview with a geophysicist from London Mining Network, September 11, 2015.

unfolded alongside the aggressive plundering and depletion of planetary ecosystems, the fourth industrial revolution Schwab describes is
essentially suboptimal and unrevolutionary.[59] This dubious technological "revolution," Moore argues, has nonetheless been able to bring about
transformations that are unrivaled in scale and scope, such as the
"conversion of the global South into a 'world farm,' the industrialization
of the south, and the radical externalization of biogeophysical costs,
giving rise to everything from cancer epidemics to global warming."[60]
Although atypical when compared with previous industrial revolutions,
the aforementioned transformations indicate that contemporary industrial technology has already summoned into existence the most
advanced and pervasive iteration of machinofacturing yet. This has
reconfigured the modes of organizing social forms of labor and the
process of social reproduction to such an extent that it would be consistent to refer to it as a *fourth machine age* on the basis of Mandel's
periodization.[61]

The modalities of functional integration across the domains of
production, circulation, and consumption of social wealth that have
been enabled by the current industrial era, then, demand more open
and exploratory theoretical engagements with the notion of the nation-
state. Indeed, the pitfalls of methodological nationalism have most
clearly manifested in the confusion of recent literature on resource

59   World-ecological revolutions have tended to be underpinned by quantum leaps
in productive technologies that deliver a rising ecological surplus. A rising ecological
surplus, for Moore, is materialized whenever a relatively modest amount of capital is
able to set into motion a very large mass of work and energy. Moore, *Capitalism in the
Web of Life*, 149.

60   Moore, *Capitalism in the Web of Life*, 162.

61   The fourth industrial revolution can be considered an offspring of the so-called
microelectronics technological revolution because it emerges from the miniaturization
and computerization enabled by innovations in semiconductor materials based on
integrated circuits (Starosta, "Outsourcing of Manufacturing"; Lüthje et al., *Silicon
Valley to Shenzhen*). Although experimentation with semiconductor materials dates
back to the late nineteenth century, it was ion implantation, a technique that emerged
from nuclear physics after the 1970s, that enabled integrated circuits to develop their
potential in full and qualitatively transform industrial technologies (Christophe Lécuyer
and David C. Brock, "From Nuclear Physics to Semiconductor Manufacturing: The
Making of Ion Implantation," *History and Technology* 25, no. 3 (2009): 193–217. To the
extent that microelectronic devices did not substantially develop until nuclear physics
became a mature science, the existence of a fourth machine age would be consistent—
both chronologically and analytically—with Mandel's periodization.

extraction, which has faced significant difficulties in interpreting the rise of China and other Asian economies through the lenses of world-systems analysis and related frameworks. In some accounts, it is far from clear whether China has acquired the status of "hegemon" or even if it continues to be a periphery/semiperiphery at all. Others argue that China has assumed the role of the United States as the new global super-power, and as a result has transformed Latin American countries into its resource peripheries.[62] For Bolinaga and Slipak, for example, the "Washington Consensus" that once subordinated Latin American econ-omies to the interests of the United States has been replaced by what they term a "Beijing Consensus": the compromise to supply cheap raw materials to China.[63] As was the case with the US, these authors point out, the Beijing Consensus is much more of an imposition than a consensus, and thus the core-periphery model continues to be repro-duced under a different façade. The political reductionism of such approaches, it is worth insisting, confuse appearance with essence and end up obfuscating the real social determinations that animate the commodity supercycle. For these reasons, the section that follows argues that technological upgrading in the extractive industries thus present us with the challenge of unthinking modern society against and beyond its existing conceptuality.

Imperialism as Political Form

Marx's original scheme for his exposition of the critique of political economy included a final volume whose central theme would be the world market. As is well known, he only managed to write three of all the projected volumes encompassed by his original 1857 outline. From the very beginning, however, Marx conceptualized capitalist society as a world system, so for him the world market was the place "in which production is posited as a totality together will all its moments."[64] It is

---

62 Luciano Bolinaga and Ariel Slipak, "El Consenso de Beijing y la reprimarización productiva de América Latina: El caso argentino," *Revista Problemas del Desarrollo* 183, no. 46 (2015): 33–58; Stefan Schmalz, "El ascenso de China en el sistema mundial: Consecuencias en la economía política de sudamérica," *Pléyade* 18 (2016): 159–92.

63 Bolinaga and Slipak, "Consenso de Beijing."

64 Cited in Braunmühl, "On the Analysis of the Bourgeois Nation State," 163.

therefore the place where total social capital becomes inverted into the alienated subject of the process of social reproduction in its unity—the realization by capital of itself as a totalizing and totalized subject.[65] The idea behind the Marxian concept of the world market is that the total surplus value of society—itself a product of myriad embodied acts of production and exchange—is determined prior to its distribution among individual capitals and states. The world market, Bonefeld argues, is the most developed form of this abstract interconnectedness.[66] This means the global economy should not be understood as an aggregation of national economies. Rather, the latter are most adequately conceptualized as the modes of existence assumed by the former, whose reality springs from the actual, material metabolism of social life.

The world market is therefore not an abstract substance or "structure" that is either transcendent or somehow extrinsic to human experience. In fact, Marx intended the idea of the world market to express the primacy of the concrete corporeality of the sum total of the vast multitude of labors living under the geographies of capitalist society, before it assumes distorted forms in international political relations and intercapitalist competition. In the introduction to the 1978 edition of volume 2 of *Capital*, Ernest Mandel argues that "it was only by dealing with the reproduction of capital *in its totality* that Marx could bring out in their full complexity the inevitable contradictions of the basic cell of capitalist wealth, the commodity."[67] One of the most insightful treatments of this aspect of the critique of political economy may be the one developed by the Argentinean philosopher Enrique Dussel in his 1988 book *Hacia un Marx desconocido: un comentario de los manuscritos del 61–63* (published in English in 2001 under the title *Towards an Unknown Marx*). Dussel attempts to elucidate the global nature of capitalism through an exploration of the recently discovered *Manuscripts of 1861–63*, where Marx reflects on the logical structure of the three volumes of *Capital* and on the centrality of the question of distribution. On the basis of Hegel's idea

---

65 Postone, *Time, Labor, and Social Domination*; Starosta, "Revisiting the New International Division of Labour Thesis"; Bonefeld, *Critical Theory*, chapter 7.

66 Bonefeld, *Critical Theory*, 152.

67 Mandel, in Marx, *Capital: A Critique of Political Economy*, Vol. 2 (New York: Penguin Books, 1992 [1978]), 12.

of the inverted world,[68] Dussel expounds how prices, rents, and profits constitute the phenomenal or sensuous manifestation of essential being, which is value. It is through the process of distribution via circulation, competition, and sale that the essential level (surplus value) acquires phenomenal reality in profit.[69] The political mediations of the interstate system would then constitute the historical or phenomenal expressions of the distribution of the total surplus value of society.

For Dussel, most of the theories of dependency that emerged throughout the 1960s in Latin America and beyond were substantially flawed because they confused the essence of dependency (transfers of surplus value) with its multiple, phenomenal, historical appearances (international political relations).[70] The approach proposed by Dussel, it is worth mentioning, shares important elements with that of other radical Latin American theorists of dependency, such as Ruy Mauro Marini and Agustín Cueva. Despite the particularities and internal nuances of each of these authors, they all strove to overcome the shortcomings that methodological nationalism brings to the study of dependency, and rooted the political dynamics of the interstate system on the exploitation of labor-power and the production of relative surplus value.[71] Marini's approach to dependency is perhaps the most influential and illustrative in this regard. For Marini, the possibility for "developed" nations to establish capital-intensive industrial systems oriented toward the production of relative surplus value was premised upon the "super-exploitation" of the laboring classes in "underdeveloped" nations, where production was organized on the basis of the extraction of absolute

68   For an overview of Hegel's notion of the inverted world as well as his organismic reading of the absolute, see Pippin, "You Can't Get There from Here"; Reichelt, "Social Reality as Appearance."

69   Dussel, *Towards an Unknown Marx*, 149. According to Marx, "the capitalist knows nothing of the essence (*wesen*) of capital, and surplus value exists in his consciousness (*bewusstsein*) only in the form of profit, a converted form of surplus value . . . In fact, the capitalist himself regards capital as a self-acting automaton, which has the quality of increasing itself and bringing in gain, not as a relation, but in its material existence" (ibid.).

70   Dussel, *Towards an Unknown Marx*, 220.

71   Dussel, *Towards an Unknown Marx*, chapter 13; Andy Higginbottom, "Underdevelopment as Super-exploitation: Marini's Political-Economic Thought," unpublished draft presented at the 2010 Historical Materialism Conference; Ariel Slipak, "Ruy Mauro Marini, un imprescindible para el debate latinoamericano," *Cuestiones de Sociología* 14 (2016), e007.

surplus value from labor-intensive production of raw materials and foodstuffs.[72]

In considering the essence of dependency to be the transfers of surplus value between "dominant" and "dependent" national economies, however, authors such as Dussel, Cueva, and Marini could not fully supersede the political reductionism or state-centrism that they set out to criticize. However, Dussel's method of developing a logical progression from the determination of the total surplus value of society—the *global* total social capital—toward the determination of its individual parts—firms and states—offers fundamental insights for making sense of the geopolitical context that leads to the contemporary organization of extraction.[73] More recently, and also on the basis of a Hegelian reinterpretation of Marx's concept of the world market, a group of Latin American scholars have revisited the thesis of the international division of labor in order to rethink some of the key assumptions and implications of dependency theory.[74] These authors are associated with the Buenos Aires-Based Centro para la Investigación como Crítica Práctica (CICP), under the direction of the Argentinean scholar Juan Iñigo Carrera. In contrast to Dussel, they suggest that the immanent content that governs the dynamics of the NIDL (its *essence*) is not dependency but the production of relative surplus value at the world scale, and the fragmentation of the productive subjectivity of the international working class.[75]

It is in the contradictory and crisis-ridden tendency to constantly revolutionize the technological basis of the forces of production where, according to Starosta, the foundation for the evolving, uneven spatial

---

72   Ruy Mauro Marini, *Dialéctica de la Dependencia* (Mexico City: Era, 1973).

73   For further insights on this approach, see Moseley, *Money and Totality*, chapter 1.

74   Iñigo Carrera, *El Capital*; Starosta, "Revisiting the New International Division of Labour Thesis"; Caligaris, "Global Accumulation of Capital."

75   The pursuit of relative surplus value is the foundation of the technological dynamism characteristic of capitalist society. When an individual capitalist improves the technical conditions of production, this provides her with an extraordinary profit. Competition, David Harvey explains, impels capitalists toward perpetual revolutions in the forces of production to such an extent that technological change becomes fetishized, assuming the status of an end in itself. David Harvey, *The Limits to Capital* (New York: Verso, 2006 [1982]), 122.

differentiation of the international division of labor should be sought.[76] Such is the general content expressed by means of the evolving political forms of national policy and international competition. The pathways of national development across East Asian economies during recent decades illustrate this. In the 1950s, the garment and clothing industries in the West offshored parts of their production to Japan. In the 1970s, as the Japanese laboring classes became more skilled and were put in charge of more complex tasks—such as microelectronics assembly, automobile manufacturing, etc.—the cheaper and less complex parts of the labor process began to be relocated to Taiwan, South Korea, Hong Kong, and Singapore. A similar process ensued with this first generation of Asian Tigers, which put pressure on capitalists to relocate tasks of low complexity yet again—and the latent surplus populations of Thailand, Malaysia, the Philippines, and Indonesia became the new frontier for capitalist commodification in the early 1980s. A further round of incorporations took place after the 1980s that included Bangladesh, Sri Lanka, and Mauritius, among some of the other states now known as "tiger-cub economies."[77]

Although China is a newcomer to this transnational and frenzied process of mass proletarianization, its incorporation in the 1990s constituted an event of world-historical significance. The social composition and sheer size of its latent surplus populations (migrant and indebted peasants), coupled with the iron fist of China's Communist Party, produced the largest industrial proletariat the world has ever seen. It is estimated that around 400 million peasants have been incorporated into the expanding constellations of China's industrial towns and cities since Deng Xiaoping's landmark 1992 southern tour speech.[78] The speed and intensity at which these contradictory and crisis-ridden tendencies are playing out in contemporary China are astonishing. As Hao Ren vividly illustrates,[79]

---

76 Starosta, "Revisiting the New International Division of Labour Thesis."

77 For an account of the temporality of the NIDL, see Hung, "Introduction."

78 Lawrence Liaw, "Genesis and Evolution of Chinese Factory Towns in the Pearl River Delta: From Hong Kong Towards Shenzhen," in Stefan Al (ed.), *Factory Towns of South China: An Illustrated Guidebook* (Hong Kong: Hong Kong University Press, 2012); Hao Ren, Zhongjin Li, and Eli Friedman (eds.), *China on Strike: Narratives of Workers' Resistance* (Chicago: Haymarket Books, 2016).

79 Ren et al., *China on Strike*, 221–22.

the landscapes of worker contestation in China have been torn asunder by simultaneous forces of industrialization and deindustrialization, as capital relocates some parts of the labor process toward the country's interior and to India and other countries. In the Pearl River Delta, for example, some workers mobilize in capital-intensive factories demanding higher wages, while others—literally a few kilometers away—protest next to abandoned factories for severance compensation and unpaid wages.[80] Local governments in industrial areas across China, according to Ren, are less interested in retaining low-end manufacturers and prefer to attract capital-intensive production. This means that the emphasis is deliberately placed on relative surplus value (i.e., fixed capital), not on absolute surplus value (i.e., variable capital/living labor).

Placing the emphasis on the system-wide production of relative surplus value, therefore, explains how the endless pursuit for ever-increasing levels of productivity gives rise to a society that, according to Postone, is directionally dynamic. Although social, Postone explains that the "treadmill effect" unleashed by the self-expansion of value becomes independent from human will and acquires an objective, lawlike quality.[81] Figure 3 illustrates this treadmill effect in the aggressive tendency toward mechanization taking place during the last two decades, especially as expressed in world exports of machinery and electrical equipment. After 2001, when China began to secure access to raw materials abroad systematically, the volume of machinery exports increased by leaps and bounds with respect to its previous variability (see also figure 4, which shows the evolution of China's imports of raw materials). Ever since the publication of *Technics and Civilization* in 1934, Rosalind Williams suggests, Lewis Mumford searched frantically for images that could express the all-encompassing nature of the expanding technological environment, of the sort encapsulated in figure 2.[82] In the 1960s, Mumford continued to reflect on the encroaching mechanization of social existence and invented the term *megatechnics* to highlight that human civilization was moving toward the

80   Ibid.
81   Ibid., 290.
82   Rosalind Williams, *Notes on the Underground: An Essay on Technology, Society and the Imagination* (Cambridge, MA: MIT Press, 2008).

development of a "uniform, all-enveloping structure,"[83] perhaps akin to the "cosmic megastructures" that theoretical physicists refer to as Dyson spheres.[84]

Figure 2 Total world exports of machinery and electrical equipment, 1988–2015

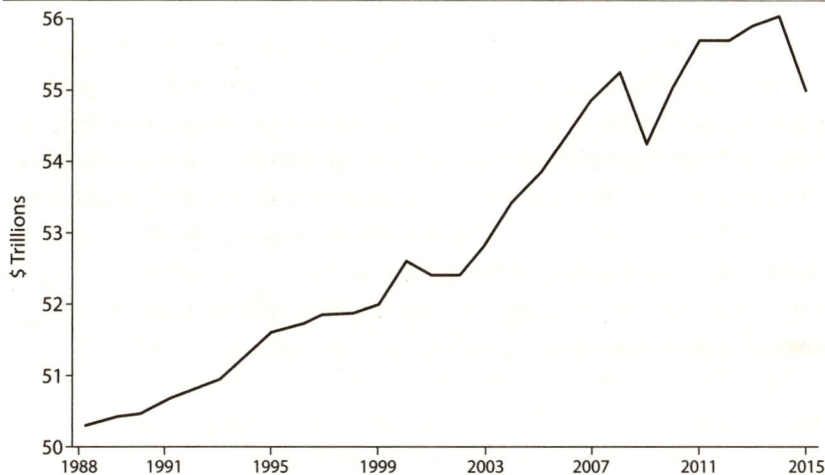

Source: Data from the World Integrated Trade Solution—WITS.

Trying to identify an "empire-building project" or a quest for "world domination" by China, any other Asian state, or the US as the true explanation for these fragmenting, self-actualizing powers would be tantamount to mistaking a symptom for the disease. The foundations of the uneven spatial differentiation of global capitalism, Starosta points out, must instead be searched for "in the changing forms of the exploitation of the global working class by the total social capital through the transformation of the material forms of the capitalist labor process."[85] To reiterate, this is the essential content of a process that acquires phenomenal reality in the sensuous materialities of corporate practices and state policy—which, of course,

---

83   Cited in Williams, *Notes on the Underground*, 7.

84   The idea of cosmic megastructures was first proposed by physicist Freeman Dyson in a 1960 paper in the journal *Science*. Dyson suggested that advanced extraterrestrial civilizations could build a gigantic shell to enclose their parent star (their sun) to collect every iota of its energetic output. Adam Hadhazy, "Could We Build a Dyson Sphere?," *Popular Mechanics*, August 19, 2014, popularmechanics.com.

85   Starosta, "Revisiting the New International Division of Labour Thesis," 86.

often present themselves as imperialist practice or regulatory strategies. The implications of this approach for understanding the shifting geographies of resource extraction in Latin America and beyond
are fundamental. According to Caligaris, the constitution of a
national sphere geared toward raw materials exports only makes
sense if cheapening the commodities it supplies results in a lower
value of labor-power exploited by the total social capital.[86] In a similar vein, Moore suggests that world ecological revolutions emerge
from the necessity of the total social capital to reduce the system-
wide cost of reproducing the working classes in order to increase
productivity.[87] Cheap energy and cheap food, according to Moore,
have been historically furthered by the application of technological
revolutions to primary-commodity production—mainly agriculture—henceforth enabling new phases of world accumulation and
capitalist development.

This does not imply that state mediations have ceased to be relevant or that the process of accumulation is contingent upon minimal state intervention, as dominant approaches to globalization
portend. In the fourth machine age, access to raw materials continues to be systematically secured by extraeconomic means. As the
next section and the next chapters will illustrate in detail, police
trucks, water cannons, surveillance cameras, tear gas, and barbed
wire continue to be constitutive elements of the landscape of extraction in postcolonial Latin America. Indigenous peoples, peasants,
and other subaltern groups continue to experience the expansion
of natural resource frontiers through a corporeal phenomenology
of imperial domination, expressed in violent expulsion, environmental racism, and everyday intimidation by police, military, and
paramilitary forces. Accordingly, understanding the alienated
political forms in which the production of relative surplus value at
the world scale is concretely actualized continues to be a crucial
element in anticapitalist thought and action. As Bonefeld rightly
suggests, the state is "the political form of market liberty."[88] The

---

86   Caligaris, "Global Accumulation of Capital."

87   Moore, *Capitalism in the Web of Life.*

88   Werner Bonefeld, "Adam Smith and Ordoliberalism: On the Political Form of
Market Liberty," *Review of International Studies* 39 (2013), 233–50; Bonefeld, *Critical
Theory.*

economy has no existence in itself, so it requires a strong institutional basis in order to actualize and *enforce* its existence through police, military, and carceral regimes. Market freedom, therefore, presupposes the political state and is premised on the state as its authority.[89]

An approach that posits the social constitution of the total surplus value of society as the prior reality upon which national economies are later determined as their moving parts, however, challenges established understandings of imperialism. As McNally argues, accounts that consider capitalist imperialism to be the result of two distinct logics of power—one capitalist and one territorial—are unable to make adequate sense of the social dynamics of impersonal power that are intrinsic to capital as the alienated subject of the historical movement of modern society.[90] In these accounts, a political body that operates in "territorial space" is juxtaposed with a capitalist class that exists in "economic space" as methodologically distinct.[91] At the heart of these differentiations between politics and economics is a dualistic reading of base and super-structure that obfuscates the essential unity of the two in the capitalist production of space.

One of the most remarkable insights that emerged from the work of authors associated with the Second International and the tradition of Monopoly Capitalism is that they were sensitive to the economic foundations that underpinned and gave momentum to the process of

---

89   Bonefeld, "Adam Smith and Ordoliberalism"; McNally, "Blood of the Commonwealth."

90   McNally, "Blood of the Commonwealth."

91   Arrighi, *Long Twentieth Century*; Meiksins Wood, *Empire of Capital*; David Harvey, *The New Imperialism* (Oxford: Oxford University Press, 2005). Harvey defines capitalist imperialism as "a contradictory fusion of 'the politics of state and empire' . . . and the 'molecular process of capital accumulation in space and time.'" The politician, Harvey argues, seeks a collective advantage and is constrained by the military apparatus of the state and also by their responsibility to a citizenry. The geographical process of capital accumulation, on the other hand, is much more diffuse and less amenable to political decision-making. The fundamental point, he concludes, is "to see the territorial and the capitalist logics of power as distinct from each other" (Harvey, *New Imperialism*, 29). Likewise, Meiksins Wood's notion of "the empire of capital" is explicitly premised on the separation of the "economic" and the "political." This "makes possible the unbounded expansion of capitalist appropriation by purely economic means and the extension of the capitalist economy far beyond the limits of the nation state" (Meiksins Wood, *Empire of Capital*, 23).

political imperialism. To cite a well-known example, Rosa Luxemburg conceptualized imperialism as the political manifestation of the process by which capital sustains its process of enlarged reproduction by appropriating the noncapitalist environment or "outside."[92] In Luxemburg's work, however, the noncapitalist outside (periphery) was construed in eminently territorial terms, basically because capitalism was considered a relatively local socionatural system at the time. Imperialism, for her, was therefore first and foremost manifested via territorial expansion through colonization, pillaging, and military incursions. If the material constitution of the world market forecloses the possibility of subsequent territorial expansion, then the production of peripheries shifts the emphasis from spatial extension to *intensification*; hence the relevance of Marx's distinction between the formal and the real subsumption of labor to capital.[93]

The production of relative surplus value in a context of real subsumption, Mezzadra and Neilson contend, opens a new perspective "on the continuous production of this constitutive outside . . . that can continue well beyond the point when territories literally lying outside the domination of capital no longer exist."[94] Under a materialist understanding of resource extraction, the periphery should therefore no longer be expressed exclusively in terms of a geographical relation; it is also eminently a social and temporal relation. This means that capital's constitutive outside can no longer be cast in terms of a straightforward North/South or West/non-West divide. Rather, cores and peripheries need to be understood as *immanent* to the capitalist production of space. The coming of age of a

---

92   Rosa Luxemburg, *The Accumulation of Capital* (New York: Routledge, 2003 [1913]).

93   In volume 1 of *Capital*, Marx develops an analytical distinction between formal and real subsumption of labor to capital under the capitalist mode of production. In formal subsumption, capital takes over an already existing labor process developed by different and more archaic modes of production (mining, sowing, knitting, etc.) involving direct exploitation of the labor of others (1021). By contrast, real subsumption involves developing a specifically capitalist mode of production (large-scale industry). This, says Marx, "not only transforms the situations of the various agents of production, it also *revolutionizes* their actual mode of labour and the real nature of the labour process as a whole" (1021). Absolute surplus value is largely produced in a context of formal subsumption, when the organic composition of capital is low, whereas relative surplus value is characteristic of the capital-intensive dynamics of real subsumption.

94   Mezzadra and Neilson, *Border as Method*, 72; Gago and Mezzadra, "Critique of the Extractive Operations of Capital."

global sphere of accumulation, however, has not banished imperialism as a political practice, as some authors have been quick to point out.[95] From its dark genesis in the transatlantic slave trade of the sixteenth century and the parliamentary enclosures of seventeenth-century England, the modern mode of production has been contingent on the exercise of political force, the locus of which has historically been—and continues to be—the capitalist state. For this reason the expansion of resource frontiers across Latin America persistently expresses itself in the fetishized form of imperialist and subimperialist practice, even though its real necessity is in the "elsewhere" of its essential content. The next section now turns to elucidating the dialectical interaction between spaces of extraction in Latin America and their concomitant geopolitical forms.

The Geopolitics of the Planetary Mine in Latin America and Beyond

Reflecting on the geological metabolism of the affluent skyline of San Francisco, Gray Brechin suggests that modern cities are technologically, philosophically, and economically the "inverted mines" of distant resource hinterlands: mineral wealth excavated from the bowels of the earth and then fixed in the urban built environment.[96] Thinking of cities as "inverted mines," then, warrants asking what sorts of spaces of extraction are behind the fantastically alien skylines of Asian megacities, some of which seem to have been transplanted directly from the cyberpunk universes of sci-fi classics such as *Blade Runner* and *Ghost in the Shell*. According to Stephen Graham, the futuristic lightscapes assembled by Shanghai's "leap into the sky" are nothing less than the aesthetic and material manifestations of "the greatest concerted construction of vertical architecture in human history."[97] That China has recently become the major importer of raw materials in the world is therefore unsurprising. When the Communist Party assumed power in 1949, China had sixty-nine

---

95 Hardt and Negri, *Empire*; Mezzadra and Neilson, *Border as Method*.

96 Brechin, cited in Bridge, "Hole World."

97 Stephen Graham, "Vertical Noir: Histories of the Future in Urban Science Fiction," *CITY* 20, no. 3 (2016): 399.

cities; today it has 658.[98] Also, it is estimated that over the next twenty years China will build hundreds of new cities, thousands of new towns and districts, and more than 50,000 new skyscrapers.[99] The country's voracious hunger for raw materials has led it to embark on a Promethean project to cast a wide net of logistical infrastructures across the seas, rivers, deserts, and mountain ranges of Asia, Africa, and Latin America. This leads Parag Khanna to argue that China "is not 'buying the world' per se but *building it* in exchange for natural resources."[100] Figure 3 illustrates the exponential increase in China's imports of minerals after the turn of the century.

Figure 3 China's imports of minerals and stone/glass, 1995–2015

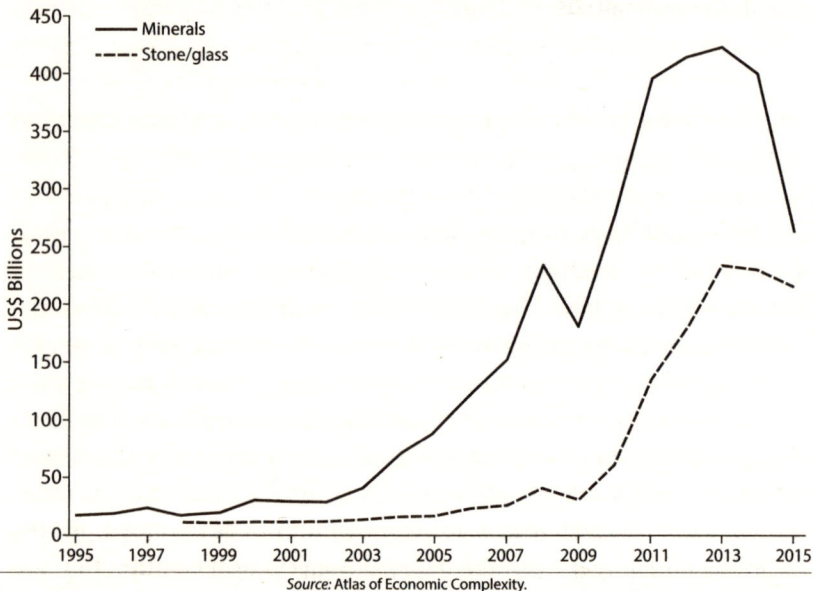

Source: Atlas of Economic Complexity.

Although Chinese investment for primary-commodity production has had a much more marked presence in Africa so far, China has already become the main commercial partner for many Latin American economies. The growing density and breadth of this trans-Pacific

---

98   Shepard, *Ghost Cities of China*, 5; Hyun Bang Shin, "Urbanization in China," *International Encyclopedia of the Social and Behavioral Sciences*, 2nd ed., vol. 24 (New York: Elsevier, 2015), 973–79.

99   Shepard, *Ghost Cities of China*, 5.

100   Khanna, *Connectography*, 95.

industrial metabolism is starkly manifested in the evolving balance of trade between China and Latin America, which has grown exponentially in recent years, going from $15 billion in 2000 to $200 billion in 2011.[101] As of 2013, South America supplied 60 percent of Chinese soybean imports, 80 percent of fishmeal, 60 percent of poultry meats, and 45 percent of grapes.[102] Just to put China's soybean consumption in global perspective: the US, Argentina, and Brazil produce 80 percent of the world's soybean yield, half of which is then exported to China.[103] These figures should not be surprising, considering that the sprawling growth of complex constellations of manufacturing towns ("township village enterprises"—TVEs) and cities in China, coupled with impoverishment and the destruction of livelihoods in the countryside, have transformed hundreds of millions of "free" peasants into wage-laborers. As the previous section suggested, the shifting modes of existence of the fragments of the global working class—paradigmatically exemplified in the social reproduction of the Chinese industrial proletariat as expressed in food consumption (see figure 4)—constitute the underlying determination for the production resource peripheries.

Paradoxically, securing the material conditions for the social reproduction of the swelling Chinese proletariat—composed mainly of displaced rural populations—has hinged upon the mass displacement of Latin American peasantries. Producing cheap food to feed the newly proletarianized populations in Asia, then, cannot be disentangled from the deforestation of the Amazon and the Chaco regions to make way for agroindustrial investment projects; from the cancer epidemics and malformations of communities destroyed by unrestrained use of glyphosate for transgenic soybean crops in the Argentinean Pampa; from the pervasive effects of the antibiotics and disease in maritime ecosystems and food chains endemic to industrialized aquaculture in Chile. The dialectic of implosion and explosion at the heart of contemporary approaches to planetary urbanization is nowhere more clearly manifested than in such relational geographies.[104]

---

101 Valdez Mingramm, "China y América Latina," 32.
102 Martínez Rivera, "China y América Latina y el Caribe," 142.
103 Trápaga Delfín, "Introducción," 167.
104 Brenner (ed.), *Implosions/Explosions*.

Figure 4 China's imports of foodstuffs, 1995–2015

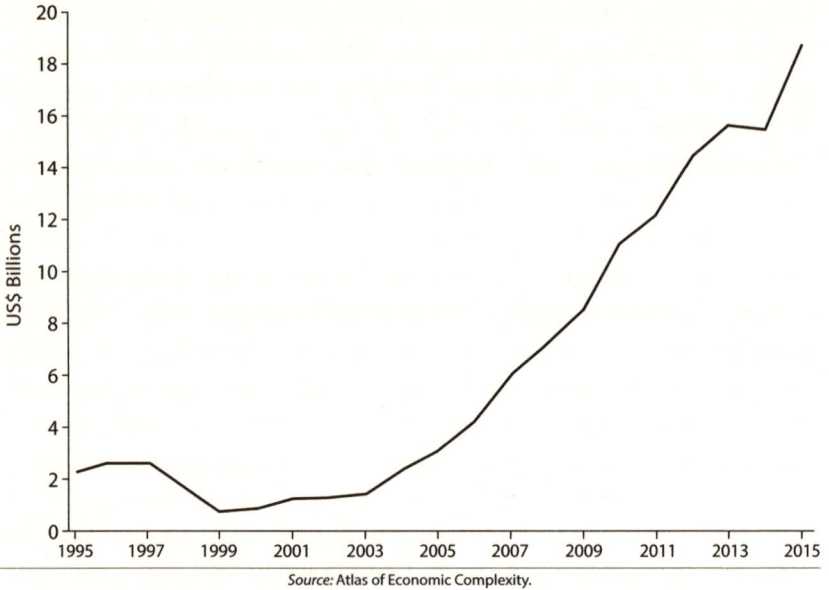

Source: Atlas of Economic Complexity.

Systemic pressures for the mining industry to become increasingly capital intensive also respond to the dramatic increase in demand that has followed the industrial expansion and urbanization taking place across East Asian economies. The tendencies toward technological upgrading, mechanization and material expansion of mining sites have had a direct repercussion on the overall volume of mineral exports. In 1990, Chile produced 16 percent of the world's copper; during the early 2000s it almost doubled its production, supplying 30 percent of world copper consumed.[105] From being almost marginal in the early 1990s, Chile's exports of raw materials to China have skyrocketed. Exports of copper, for example, increased from $3.9 billion in 2005 to $14.6 billion in 2012, and exports of cellulose rose from $335 million to $901 million during the same time period.[106] China consumes 40 percent of world copper production, so this has implied further trade with Chile as well as direct acquisition of mines in order

---

105   Fundación Chile, *Desde el cobre*, 65.

106   DIRECON—Dirección General de Relaciones Económicas Internacionales, *Evaluación de las relaciones comerciales entre Chile y China a siete años de la entrada en vigencia del tratado de libre comercio* (Santiago: Ministry of Foreign Affairs, Government of Chile, 2013), 7.

to cope with China's domestic demand, which has grown in tandem with the electronics, alternative energies, and automobile industries.[107] In Peru, the expansion of the mineral extraction frontier has been particularly haphazard and frantic. Between 2002 and 2008, the area of mining concessions in Peru increased by a staggering 77.4 percent.[108] Between 2004 and 2008, the proportion of Peru's Amazon basin covered by hydrocarbon concessions went from 14 to more than 70 percent.[109]

The expansion of iron-ore mines across Latin America is also directly connected with the booming steel industry in China. Building on the computerized technologies for iron-ore smelting first pioneered by Japan, Chinese steel production increased from 40 million tons in 1980, to 489 million tons in 2007, a figure that accounted for 36 percent of world steel production. As of 2001, China had already become the world's leading steel producer, and one of the main iron-ore importers.[110] As a result, the acquisition and growth of iron-ore mines across Latin America (mainly Brazil and Peru) has also increased substantially. The mounting intensity and density of flows in the transpacific logistical corridor of mineral trade is reflected in China's agreement with Vale—Brazil's flagship mining company—to engineer and build the *Valemax*, the largest bulk carrier ship in the world. With a capacity of 450,000 deadweight tons (dwt), this mammoth vessel carries Chinese coal to Brazil and Brazilian iron ore back to China.[111]

Although these sociometabolic interdependencies hint at the fact that China could potentially fit into the category of a hegemon or ascending power, a closer look reveals a much more complex reality than a mere quest for global dominance. As opposed to Western

---

107   Henry Sanderson, "China Wants to Buy More Copper Mines to Secure Its Supply," *Financial Times,* April 6, 2016, ft.com.

108   Anthony Bebbington, Denise Humphreys Bebbington, Leonith Hinojosa, María Luisa Burneo, and Jeffrey Bury, "Anatomies of Conflict: Social Mobilization and New Political Ecologies of the Andes," in Anthony Bebbington and Jeffrey Bury (eds.), *Subterranean Struggles: New Dynamics of Mining, Oil, and Gas in Latin America* (Austin: University of Texas Press, 2013), 245.

109   Bebbington et al., "Anatomies of Conflict," 245.

110   Bunker and Ciccantell, "Economic Ascent of China" and *Globalization and the Race for Resources*; Ciccantell, "China's Economic Ascent."

111   Bunker and Ciccantell, "Economic Ascent of China," 579.

colonial powers, Khanna argues that China is not interested in occupying territories, let alone direct political intervention.[112] Instead, it seeks to ease passage across them and steer the direction of raw material flows. As a result, Khanna asserts, the global presence of China is defined not so much by its military forces as by its supply chains. By emphasizing the transformative role of connectivity infrastructure in the everyday practice of populaces and political bodies, Khanna points toward the fact that the production of relative surplus value is a process of a higher ontological order than the political mediations of the nation-state. In an illustrative passage, Khanna asserts that "for China, supply chain blowback *is* geopolitical blowback."[113] It is by riding the wave of material-technological expansion that China, according to Khanna, has been able to build a "global supply-chain empire" without needing the geopolitical maneuvers of old forms of imperialism.

The predominant view among scholars in the fields of international relations and Latin American studies is that China's rise as a commercial partner for Latin American economies marks a new paradigm of nonhegemonic, multipolar, "cooperative" international relations.[114] This emerging geopolitical framework, so the argument goes, contrasts starkly with the vertical and militaristic relations of the region with the American, British, Dutch, and Iberian empires. Whereas British and US relations with Latin America were interpreted as those of the enlightened master lending a charitable hand to a disciple, China is often construed as more of an ally that seeks reciprocal benefits. In terms of natural resources, China has sought diplomatic-strategic alliances with twenty-one of thirty-three Latin American nation-states; unlike the United States, it has expressed clear intentions to accept more equitable profit rates as well as to implement strict noninterference policies.[115] More than "assistance," commercial relations between China and Latin America have been characterized as relations of complementarity and South-South cooperation.[116]

Chinese investment strategies for primary-commodity production are also markedly different from those of other economic powers such

112   Bunker and Ciccantell, *Globalization*.
113   Khanna, *Connectography*, 221.
114   Napoleoni, "La política china"; Diana Sofía Báez Pichucho, "Ecuador y China, socios petroleros," in Trápaga Delfín (ed.), *América Latina y el Caribe-China*.
115   Báez Pichucho, "Ecuador y China."
116   Ibid.

as Canada, Britain, and the United States. Whereas China appears to be wholeheartedly interested in furthering transfers of technology and know-how in order to improve the technical conditions of extraction with the strategic partner, the US limits its participation to mere exploitation of mineral deposits or oil reserves.[117] Perhaps what is most revealing about China's bilateral relations with Latin American states is that Chinese leadership is decidedly framed in commercial rather than military terms.[118] Chinese diplomatic strategies to procure access to natural resources in Africa have been said to follow a very similar logic, because China's actors adapt to the particular histories and geographies of the African states with which they engage. Aptly termed *flexigemony* by Carmody and Taylor,[119] the incipient forms of rule and modes of natural-resource governance introduced by China mark a clear break with the neocolonial and realist orientations of Western economic powers. Whereas neoliberal "virtualism" sought to make the actuality of social relations conform to an ideal type, Chinese interests build upon the principle of peaceful rise (*heping juequi*) to dictate a rhizome-like approach that uses existing networks of influence but also creates new ones.[120]

These incipient modes of geoeconomic resource governance, however, are far from unproblematic or peaceful. In the style of the narrative twists of a Hitchcock film, where the macabre is unexpectedly superimposed on the idyllic, unsettling visions emerge as one descends from the nonhegemonic, harmonious diplomatic alliances of China to the everyday spatiotemporality of extraction mediated by them. Kidnappings and attacks against Chinese engineers and workers in the primary sector, according to Khanna,[121] are on the rise from the Niger Delta to southern Sudan. Zambian miners, he argues, "have violently rebelled against their Chinese employers' slave wages and slave-driving tactics, on several occasions trampling, crushing, and killing them deep inside mine shafts."[122] Mounting levels of police violence, state crackdowns, and even genocidal

---

117 Ibid.
118 Napoleoni, "La política china."
119 Carmody and Taylor, "Flexigemony and Force"; Carmody, *New Scramble for Africa*.
120 Carmody and Taylor, "Flexigemony and Force."
121 Khanna, *Connectography*, 222.
122 Ibid.

war have become part and parcel of the experiential basis of some of the expanding territories of extraction of this new geopolitical reality. The gold, gas, uranium, and oil deposits of the Baluchistan province in Pakistan, to cite an example, have led to the fierce suppression of local communities at the hands of the Pakistani army and Chinese state-owned mining companies.[123] The Baluchistan Liberation Army has retaliated by sabotaging pipelines, blowing up crowded buses, and killing numerous Chinese engineers near the infamous port of Gwadar.[124] Mines of cobalt and tantalum in the Democratic Republic of Congo, notoriously connected to the supply chains of the giant global contractors of the electronics industry—the majority of which are located in Asia—have also been at the center of international controversy due to the encroaching presence of slave labor, child soldiers, mass rape, and genocide.[125]

As Ho-fung Hung argues, gone are the days when activists attributed many of the social and political ills of the developing world to Washington or Washington-based international financial institutions.[126] To add to Hung, it is worth pointing out that activists (in Latin America, at least) do not see China as a direct imperialist power either. The paradox of it all is that the experience of imperialist practice continues to exert a powerful imprint upon the everyday lives of the local communities that coexist with infrastructure and resource-extraction projects. A 2013 manifesto of feminist activists mobilizing against resource extraction frames this context as follows:

> We are Latin American women and our identity was forged in the resistance to the colonial conquest of our territories and the pillaging of our land's commons. After more than five centuries, we continue to face ever-renewed forms of colonialism and patriarchy, now at the hands of transnational corporations who, backed by national governments, plunder and steal our common goods, thus moving forward with the silent genocide of our people.[127]

---

123 Ibid., 225.
124 Ibid.
125 Merchant, *One Device*, chapter 2.
126 Hung, "Introduction."
127 Observatory for Mining Conflicts in Latin America (OCMAL), *Feminist Manifesto against Large-Scale Mining and the Extractivist-Patriarchal-Colonial Model*, 2013, ocmal.org.

The underlying content of global capital accumulation, therefore, often presents itself in contemporary spaces of extraction under the guise of a "banal neoimperialism."[128] A whole material culture of institutional practice and technical artifacts—objectified, for example, in police raids and harassment, security fences, trenchant pollution, swarms of security guards, pamphlets containing death threats, and judicial orders of incarceration, among many others—weaves an everyday imperial reality that reproduces the same modalities of extraeconomic force that have marked Latin American history for centuries. A 2018 report by Global Witness found that 2017 was the most dangerous year on record for defenders of the environment, as 207 indigenous, land, and environmental activists were murdered.[129]

Behind such mystified political forms is not an internally determined "hegemon" that operates in territorial space and that consciously and autonomously articulates a geopolitical project. The sheer magnitude of the means of production conjured by industrial upgrading in China and other Tigers has set into motion destructive forces that seem to lie beyond any form of collective mediation, whether from state or market. A single open-cast mine can be operated by up to a dozen companies from different countries, performing different functions with different workforces, some of which are often produced by transnational migratory flows and not even under a direct work relation but outsourced and subcontracted. As we will see in the following chapter, open-cast mining has become so capital-intensive, as well as racialized and gendered, that new and variegated actors have been drawn into geographies of extraction, typically operating alongside mining corporations in ways that hamper transparency and accountability. When a mine is directly operated by a Chinese corporation (state-owned or otherwise), the complex and pervasive socioecological effects are no different from properties controlled by Western companies notorious for human rights abuses and socioecological degradation.[130] What is distressing about

128   Steven Flusty, Jaeon Dittmer, Emily Gilbert, and Merje Kuus, "Interventions in Banal Neoimperialism," *Political Geography* 27 (2008): 617–29.

129   Global Witness, "At What Cost? Irresponsible Businesses and the Murder of Land and Environmental Defenders in 2017," 2017, globalwitness.org.

130   Wilson and Bayón, "Concrete Jungle"; Japhy Wilson, "The Village That Turned to Gold: A Parable of Philanthrocapitalism," *Development and Change* 47, no. 1 (2015): 3–28;

the forces of destruction being unleashed is that rarely can a single actor, political or economic, be held accountable for them.

The "empire of muddle," which Lewis Mumford described in making sense of the disintegrating forces brought about by the mechanization of resource extraction during the first industrial revolution,[131] seems to have become much more advanced and systematic in the present industrial age—the era that Klaus Schwab and others paradoxically celebrate as the pinnacle of human progress. As Marx and Engels prophetically admonished in the *Communist Manifesto*, modern bourgeois society resembles "the sorcerer, who is no longer able to control the powers of the nether world whom he has called up by his spells."[132] Spatial technologies of extraction are traversed by violence, dispossession, and ecological destruction. Its prime mover, however, is not an autonomous imperial power but the abstract, directionally purposed forms of social mediation that assert and reassert capital as the alienated subject of modern life. In an interview, a planning official of a mining town in Chile remarked that poisoned rivers, encroaching local unemployment, air pollution, and cancer epidemics are clearly connected to mining and energy megaprojects, but accountability becomes elusive when layers upon layers of contractors, companies, outsourced workers, and subsidiaries operate simultaneously. He juxtaposed this with state-developmentalist regimes of resource extraction of previous decades, where there was far less complexity in the technical division of labor and the mining company became directly embedded in the life of the community.[133]

In this sense, perhaps nothing reflects more clearly the actual processes at work in the geographies of large-scale mining than what Pádraig Carmody has labeled "the new scramble for Africa."[134] Against ideological visions that posit external neocolonial penetration into the continent to reap its natural resources, Carmody illustrates how

---

Rubén González-Vicente, "South-South Relations under World Market Capitalism: The State and the Elusive Promise of National Development in the China-Ecuador Resource Development Nexus," *Review of International Political Economy* 24, no. 5 (2017): 881–903.

131   Mumford, *Technics and Civilization*, 194–95.

132   Karl Marx and Friedrich Engels, *The Communist Manifesto* (New York: Penguin Books, 2002 [1848]), 70.

133   Interview with an official from Vallenar's Planning Department (SecPlan), December 3, 2013.

134   Carmody, *New Scramble for Africa*.

numerous alliances between the BRICS and BRICS-based companies have developed coordinated modes of engagement with African countries that do not resemble traditional core-periphery models. The most important C in the BRICS, according to Carmody, may not be China but capitalism.[135] The big C in the BRICS, Carmody concludes, is global and operates according to its own laws, which are personified by a very wide array of actors and institutions.

Conclusion

In this chapter, I have argued that the coming of age of the planetary mine brings with it the pressing need to conceptualize capitalist society as an organic whole, not as an aggregation of national economies. The industrialization of the global South, coupled with the technological and industrial upgrading taking place across East Asia, requires a conceptual apparatus that can capture the transformation of capitalism into a genuinely global—not merely Western—form of social mediation. Accounts that seek to understand spaces of extraction *exclusively* in terms of the international political relations of the nation-state (in the guise of dependency, unequal exchange, core-periphery dynamics, and so forth), I have argued, confuse the essence of global capitalism with its multiple, historical, and phenomenal appearances. By pointing to the need to consider the political mediations of the nation-state as the modes of existence of a prior reality— i.e., the production of the total surplus value of society—this chapter has provided some methodological elements for taking seriously the differentiated, yet unitary nature of politics and economics. Processing social relations into particularized political categories is a constant struggle to suppress the expression of class experience and to transform class relations into nonclass forms.[136]

Although this discussion may appear somewhat esoteric and abstract, it has definitive political implications. Reducing

---

135   Pádraig Carmody, "The New Scramble for Africa," *Jacobin* 19, 2015, jacobinmag.com.

136   John Holloway and Sol Picciotto, "Capital, Crisis and the State," in Clarke (ed.), *State Debate*; Susan Ferguson and David McNally, "Precarious Migrants: Gender, Race and the Social Reproduction of a Global Working Class," *Socialist Register* (2015): 1–23.

natural-resource governance to a political interaction between "cores" and "peripheries" embodied in abstractly constituted nation-states is tantamount to accepting the fetishization of class struggle into distinct political and economic channels. Most importantly, it leads to the sort of reformist view where it would be considered possible to transform society by the mere conquest of political institutions. The case of the postneoliberal governments of Latin America's "Pink Tide" illustrates that claims to "national liberation" are severely limited without a more comprehensive project to supersede the modern forms of labor that act as the foundation of the political apparatus. Numerous studies have demonstrated that despite their intention to overturn the hierarchical relations of the interstate system by political means, postneoliberal governments in Latin America became even more dependent on primary-commodity exports and more aggressively subsumed in the cyclical compulsions of the world market.[137] To reiterate, this does not mean that the left should abandon anti-imperialism as a political idiom. Anti-imperialist struggle can fulfill (and very often does fulfill) an important political role, but when strategically mobilized as an entry point to more directly challenge the historical thrust of class domination.

Also, the concept of the fourth machine age developed in this chapter warrants a caveat. As Bonefeld et al., rightly stress, periodizations are to be approached with caution.[138] Dividing the history of the modern mode of production into periods can be insightful only insofar as this allows us to better understand the continuity of the movement of contradiction constituted by the reproduction of class relations. For this reason, more than an issue of semantics and of fetishizing "newness," framing our technological present in terms of a *fourth machine age* should serve as a heuristic to ask the really important questions: What does the current overhaul in the sociotechnical basis of large-scale industry reveal about the nature of state power, of capital accumulation, and of the class agencies that animate modern society? Most importantly, framing our present era in terms of a fourth machine age, in my view, should also be

---

137   Gudynas, "Agropecuaria"; "Value, Growth, Development"; Thomas Purcell, "'Post-Neoliberalism' in the International Division of Labor: The Divergent Cases of Ecuador and Venezuela," in Charnock and Starosta (eds.), *New International Division of Labor*; Wilson and Bayón, "Concrete Jungle."

138   Bonefeld, Gunn, and Psychopedis, "Introduction," in *Open Marxism*.

considered a political *dispositif* or critical idiom that departs from the liberal boosterism and circulationism of the mainstream literature, which tends to be unaware of the crisis tendencies intrinsic to late capitalist technology. Pundits tend to present this "fourth industrial revolution" in terms of a horizon of possibility or peril, as if its effects were yet to materialize in some unspecified future. One only needs to scratch the surface to find out that the alleged future horizons of the fourth industrial revolution are already at work in the dark undersides of contemporary industrial capitalism.

Finally, this chapter has also demonstrated that an approach that makes analytical distinctions between the essential movement of value and its sensuous political manifestations is also particularly useful in making sense of the pervasive presence of imperialist practice amid an alleged postcolonial international context. Despite the new frameworks of cooperative and nonhegemonic resource diplomacy characteristic of the new international division of labor, indigenous peoples, women, peasants, and other subaltern groups across Latin America (and beyond) continue to experience the same logics of racialized violence, patriarchal domination, and militarization that distinguished the Western powers' strategies for securing access to raw materials. These forms of imperialist and subimperialist practice, I have argued, need to be understood as a fetishized expression of a deeper underlying content. Perhaps *the* most salient contribution of Marx's critique of political economy consisted in "the treatment of surplus-value independently of its particular forms as profit, interest, rent, etc."[139] Classical political economists, in Marx's view, failed to get at the taproot of this problem; their analyses remained tethered to the phenomenal expressions of surplus value in the distorted forms of profit and rent.[140] Imperialism should therefore be understood as the experiential basis that underpins the subsumption of the constitutive outside of capital to the process of accumulation. If, as Stuart Hall once argued, race is the modality in which class is lived,[141] then imperialism should be understood as the corporeal phenomenology of a

---

139   Cited in Moseley, *Money and Totality*, 42.
140   Moseley, *Money and Totality*, 41.
141   Jim Glassman, "Critical Geography II: Articulating Race and Radical Politics," *Progress in Human Geography* 34, no. 4 (2010): 506–12.

process whose essential content is to be found in the *telos* of the value form: the production of the free market through the exploitation of the laboring class as a pulsing, breathing, planetary organism. The next chapter is devoted to understanding this latter, living moment within the metabolism of extraction.

3. LABOR

Bodies of Extraction and the Making of Urban Environments

The mine is only man,
the mineral doesn't emerge from the earth,
it leaves the human chest,
there one touches the dead forest,
the arteries of the suspended volcano,
the vein is identified, perforation takes place and dynamite explodes,
the rock is spilled, purified:
copper is being born.
No one will know how to differentiate it from the mother rock[1]

Introduction

Chile is the largest supplier of copper in the world, accounting for 40 percent of total world exports.[2] It also hosts some of the largest deposits of lithium, which means that much of the battery power that underpins contemporary urban life originates in the dry landscapes of its Andean plateaus. The bulk of mineral extraction is performed through

---

1   Neruda, *All the Odes*, 168–69.
2   Martín Arias, Miguel Atienza, and Jean Cadematori, "Large Mining Enterprises and Regional Development in Chile: Between the Enclave and Cluster," *Journal of Economic Geography* 14, no. 1 (2013): 82.

large-scale open-cast mining and is located in the northern part of the country, especially alongside the biogeographical region structured around the Atacama Desert. Antofagasta is the main economic hub in the urban system of this export-oriented region. The nearby copper mines produce 30 percent of Chile's exports, and the lithium pools and refineries located in the adjacent Salar del Carmen produce 130 tons of lithium carbonate every day. This is approximately 48,000 tons of lithium a year, enough to produce about 43 billion iPhones.[3] Due to the scale of mineral extraction, Antofagasta has four international ports operating at the cutting edge of cargo-handling technology, with almost exclusive dedication to the mining industry.

Resembling a less overblown version of the dystopian urban universes of Ai Weiwei's *The Sand Storm* and George Miller's *Mad Max*, the technological infrastructures that constantly pump minerals into Antofagasta's ports crisscross a bewildering, arid, and fractured urban landscape where wealth and destitution coexist side by side. Because 96 percent of Antofagasta's exports are minerals,[4] the cityscape has been aggressively splintered between the zones that most directly partake in the supply chain of extraction and those that aspire to do so but have been left behind. The wealthy quarters of the city are interspersed with shopping malls and opulent residential units boasting swimming pools, tennis courts, and expensive cars. Opposing the small islands of gated communities and conspicuous consumption tailored for the steady-salaried workers of the mining industry are Antofagasta's dark peripheries of informal, precarious subcontracted work. Mounting income gaps resulting from the capitalization of the mining industry and the increasing complexity in the technical division of labor have pushed most of the local population toward the impoverished outskirts of this present-day Potosí.

Faced with soaring rents, a considerable part of these displaced communities has been left with no choice but to inhabit the city's swelling shantytowns and precarious settlements, informally referred to as *campamentos*.[5] The high

---

3   Merchant, *One Device*, 119.

4   Arias et al., "Large Mining Enterprises," 82.

5   Emilio Thodes Miranda, "Segregación socioespacial en ciudades mineras: el caso de Antofagasta, Chile," *Notas de Población* 102 (2016), 203–27; Techo Para Chile, "Catastro de Campamentos 2016: El número de familias en campamentos no deja de aumentar," September 2016, techo.org/paises/chile/wp-content/uploads/2016/09/Catastro-Nacional-de-Campamentos-2016.pdf, 60.

salaries offered by the mining industry, however, have continued to attract the deprived underclasses from other parts of the country and Latin America, especially Bolivia, Peru, Ecuador, and Colombia.[6] Inflows of cheap migrant labor have bred new racial and economic anxieties among an already strained labor market and have been instrumental in differentiating workers by attributes such as citizenship, gender, and race. More than a direct result of the circling flows of mineral wealth, these fractured and polarized spaces of urbanization are better understood as manifestations of the modalities of the commodification of labor-power, inscribed into the bodies of the city's inhabitants and its buildings, streets, and infrastructures.

Far from being an anomalous sociospatial formation or an urban typology specific to primary-commodity production, Antofagasta is a veritable microcosm that encapsulates recent reconfigurations in the sociotechnical composition of the global working class brought about by the technological change and concentration of social wealth that are intrinsic to the current industrial era. However, the shifting geographies of labor that have followed rising levels of automation in primary-commodity production are far from signaling a shift toward a "post-work" society and therefore toward the dissolution of the working class, as Merrifield—revisiting Lefebvre's 1972 work *La pensée marxiste et la ville*—surmises.[7] In this sense, the aim of this chapter is to foreground

---

6 Nanette Liberona Concha, "La frontera cedazo y el desierto como aliado: practicas institucionales racistas en el ingreso a Chile," *Polis* 42 (2015): 2–15; María Margarita Echeverri, "Otredad racializada en la migración forzada de afrocolombianos a Antofagasta," *Nómadas* 45 (October 2016): 91–103.

7 Andy Merrifield, "The Planetary Urbanization of Non-Work," *CITY* 17, no. 1 (2013): 20–36. In fact, the idea of a postwork society is premised on a mechanistic, direct correlation between automation and job destruction, where the former is seen as invariably triggering the latter. Among certain variants of the academic left, the idea of the "postwork society" has been central to a rubric of utopian (and even Promethean) socialism, where machines or other technological systems perform the bulk of the work required for the reproduction of society and humans have more time to engage in "free activity" (for an overview, see Peter Frase, "Post-Work: A Guide for the Perplexed," *Jacobin*, 2013, *jacobinmag.com*; for a recent, more systematic intervention, see Nick Srnicek and Alex Williams, *Inventing the Future: Postcapitalism and a World without Work* (New York: Verso, 2015). Lefebvre develops a spatialized version of the post-work hypothesis by claiming, in a somewhat enigmatic tone, that the acceleration of automation would bring about the simultaneous demise of the city and of work: Henri Lefebvre, *Marxist Thought and the City* (Minneapolis: University of Minnesota Press, 2016 [1972]). Merrifield builds upon this Lefebvrian insight to suggest that unemployment is structurally inseparable from the dynamic of urbanization and its expansion on a planetary scale.

the (continued) centrality of living labor for the capitalist production of space in modern society. With this, I also seek to problematize the distinction between the extraction of minerals from inorganic natures and the extraction of surplus value from the human body in the labor process. According to Mbembe, the transformation of formerly free peasants into *bodies of extraction*—that is, bodies that extract minerals and are also rendered into living deposits for the extraction of value— has been an intrinsic element of capitalism since its dark genesis in the trans-Atlantic slave trade.[8]

By highlighting the role of the human bodies in the process of extraction, this chapter argues that the historical specificity of modern mining is premised first and foremost on the organization, engineering, and even reinvention of *human productive subjectivity*.[9] Using recent transformations in the mining industry as an analytical entry point, and from a value-theoretical perspective, my aim is to illustrate how the increasing organic composition of capital entails a simultaneous tendency to reduce the number of workers involved in productive labor (that is, the labor that produces surplus value) and to expand the number involved in unproductive labor (that is, the labor involved in the realization of value). As Maria Mies notes, the invisible underbelly of the capitalist wage system includes all those forms of unwaged work performed by women, subsistence farmers, indentured workers, slaves, and relative surplus populations. Far from being external to it—or "postwork"—as usually considered, these expanding constellations of unproductive work (from a value-theoretical perspective, that is) are in fact the foundation upon which the wage system as a whole is erected and made possible.[10]

Perhaps one of the most overlooked yet generative aspects of Neil Smith's production of nature thesis is that he considers the production

---

8 Achille Mbembe, *Crítica de la razón negra: ensayo sobre el racismo contemoráneo* (Barcelona: Futuro Anterior Ediciones, 2016).

9 The term *productive subjectivity* is aimed at capturing the twofold dimension of labor-power as it becomes embodied by the body of its living bearer. On the one hand, the productive attributes of workers entail the strictly material or technical characteristics required by the complexity and particularity of the tasks performed. On the other hand, productive subjectivity also concerns the "moral attributes," understood as the general forms of consciousness and self-understanding that make specific workers suitable for specific types of productive activity. Charnock and Starosta (eds.), *New International Division of Labour*, 19.

10 Mies, *Patriarchy and Accumulation*.

of nature to include the configurations assumed by human conscious-ness in the labor process.[11] The production of nature is henceforth also fundamentally concerned with the production and even reinvention of the human *qua* productive subject. Because labor is the "melting pot" where the unity between human and nonhuman natures is concretely realized, the question of living labor is fundamental for fully grasping the production and creative destruction of social space under capital-ism. It was Marx's profoundly relational and post-Cartesian reading of social reality that put living labor at the center of the metabolism of modern society by claiming that through labor, the individual "acts upon external nature and changes it, and in this way . . . simultaneously changes his own nature."[12]

The reorganization of the mining industry into global supply chains, I will therefore argue, is contingent upon the production of a polarizing, alienated industrial organism—the collective laborer—whose material constitution is contingent upon the uneven distribution of productive attributes among its organs. The existence of this sociomaterial form of life is premised on the simultaneous expansion of the productive subjec-tivity of the workers in charge of the more complex parts of the labor process (i.e., engineers, scientists, geologists, financiers), and the degra-dation of those workers who act as mere appendages of technological infrastructures or perform manual/low-skill tasks in the expanding "service economies" gravitating around mining firms (i.e., street vend-ing, gardening, catering, prostitution). The consolidation of the latter group, the chapter will illustrate, depends upon the systematic assault on rural, communitarian, and agrarian forms of sociality that have been associated with the commodity supercycle—a phenomenon that is part and parcel of the generalized sociospatial push factor that scholars of agrarian studies have referred to as "global depeasantization."[13] Building

---

11 Smith, *Uneven Development*; Neil Smith, "Nature as Accumulation Strategy," *Socialist Register* 43 (2007): 1–21.

12 Marx, *Capital*, Vol. 1, 268.

13 Farshad Araghi, "Global Depeasantization, 1945–1990." *Sociological Quarterly* 36, no. 2 (1995): 337–68; Farshad Araghi, "Accumulation by Displacement: Global Enclosures, Food Crisis, and the Ecological Contradictions of Capitalism," *Review* 32, no. 1 (2009), 113–46; Philip McMichael, "Peasant Prospects in the Neoliberal Age," *New Political Economy* 11, no. 3 (2006), 407–18; Eric Vanhaute, "Peasants, Peasantries, and (De)peasantization in the Capitalist World-System," in Salvatore Babones and Christopher Chase-Dunn (eds.), *Routledge Handbook of World-Systems Analysis* (New

upon recent reinterpretations of the Marxian notion of the collective laborer,[14] I argue that the productive capacity that results from this multiscalar, heterogeneous, and transnational industrial organism as a working whole is not consciously regulated by any of its organs, but rather seized and inverted into the powers of an alien system of social domination.

By placing the question of labor exploitation at the center of a theory of uneven geographical development, my intention is to offer a counterpoint to the externalist ideologies of nature that Smith considered to be so pervasive even in critical and environmentally minded social theory.[15] I also demonstrate that, far from involving the dissolution of the global working class, the technological change and industrial expansion triggered by the current industrial era have entailed a degree of proletarianization and of the real subsumption of humanity to capital that is without precedent in human history. The spatial manifestation of this process has been a fragmenting, expanding fabric of urbanization where wealthy global cities, technopolises, and a wildly expanding constellation of shantytowns become tightly intermingled by myriad metabolic flows. A precise statistical measurement of the size of the global collective laborer would of course be impossible, given the dynamism that is intrinsic to the notion of class. However, Ferguson and McNally claim that in rough figures, the global working class has grown by at least two-thirds (and possibly even doubled) across the neoliberal period, from something

---

York: Routledge, 2012); Cristóbal Kay, "La transformación neoliberal del mundo rural: procesos de concentración de la tierra y del capital y la intensificación de la precariedad del trabajo," *Revista Latinoamericana de Estudios Rurales* 1, no. 1 (2016), 1–26.

14   Postone, *Time, Labor, and Social Domination*; Iñigo Carrera, *El Capital*; Starosta, "Revisiting the New International Division of Labour Thesis."

15   Smith, *Uneven Development*. An externalist conception of nature, according to Smith, involves considering the human life-process to be ontologically different to that of extrahuman natures. A truly dialectical, post-Cartesian approach to nature then ought to include the human with the nonhuman in nature: a differentiated unity, but a unity nonetheless. The human species would be one among many others in the dynamic, variegated totality of nature. Urban studies and critical geographical scholarship, even in their Marxist/*marxisant* variants, have largely overlooked the central role of the changing conditions in the materiality of the labor process for the ongoing production and reproduction of urban socionatural worlds. The analytical focus is usually placed on the fixed components of capital (i.e., infrastructures, natural resources, built environments, systems of machinery), while the true alchemic force, the source of all surplus value—and hence of the inner motion of all social wealth—is ignored.

between 1.5 billion relying on selling their labor-power in 1980 to 3 billion in 2015—with more than half of this number making up the global reserve army.[16] The next section, then, starts by developing some conceptual and methodological tools to make sense of the territorial basis of these processes of commodification of labor-power.

The Collective Laborer and the Production of Nature

According to Neil Smith, with the development of industrial technologies for capitalist production, the material unity of society and nature is reproduced in a more advanced form than ever before. Hence, with the generalization of commodity production and exchange relations, previously isolated individuals and geographies become knitted together into a complex social whole.[17] As a result of its intrinsically self-expansive nature, capital has an immanent tendency to revolutionize the material conditions of social production and encompass ever wider extensions of geographical space. Insofar as the process of revolutionizing the forces of production is necessarily the outcome of heightened social cooperation, van der Pijl notes how capitalist society develops under two contradictory aspects: social inequality resulting from the projection of commodification frontiers across space, and planned interdependence or socialization of labor (*Vergesellschaftung*).[18] Capitalist expansion is, from the very outset, contingent upon the drives for cooperation and association that are intrinsic to the human species.

If the production of consciousness is an integral part of the production of general material life, as Smith is right to suggest, then the production of workers and their specific subjective formations, are a precondition for their socialization. Although work and the production of workers is not specific to modern society, it is important to highlight that, in the *Paris Manuscripts*, the young Marx considers the specificity of productive activity under capitalism to emanate from the *alienated* character of labor. Since she sells her only commodity (labor-power) to

---

16 Ferguson and McNally, "Precarious Migrants," 9.
17 Smith, *Uneven Development*, 65.
18 Kees van der Pijl, "International Relations and Capitalist Discipline," in Albritton et al. (eds.), *Phases of Capitalist Development*.

the capitalist in order to reproduce her material life, the laborer does not own the product of her work and is instead related to it as an alien object. This means not only that labor becomes an external existence to the laborer but that it exists outside of her and begins to confront her as an extraneous and antagonistic power. For this reason, and as Starosta notes,[19] the reproduction of the human life process under capitalism is a concrete form of an essentially inverted form of existence where the object dominates the subject. As these forms of alienation of the species-powers of the human evolve, Postone explains, workers are subsumed into capital, *becoming a particular mode of its existence.*[20]

The capital form hinges upon appropriating the vital capacities and potentialities of individual workers and the collective power that results from their socialization. When workers cooperate in accordance with a plan, Marx illustrates in volume 1 of *Capital*, the productive powers of the individual increase and her "animal spirits" are stimulated, but there also results "the creation of a new productive power, which is intrinsically a collective one."[21] With the progressive unfolding of these forms of cooperative interaction, Marx continues, the worker "strips off the fetters of his individuality and develops the capability of his species."[22] The result, Postone notes, is the creation of the collective laborer, a "machine" of sorts formed out of the combination of specialized individual workers.[23] The seizure or appropriation of the collective power that emerges from cooperation between individuals, however, is not historically specific. Through the notion of the *megamachine*, Mumford shows that the archetypal principles of automation later consolidated in the onset of the first industrial revolution were introduced in ancient societies as a means to enhance and systematize humans' work. The application of precise measurement, compulsive regularity, and abstract scientific calculations to organize human exertion, Mumford notes, enabled the colossal works of engineering that marked the Pyramid Age in Egypt and Mesopotamia.[24]

---

19   Starosta, *Marx's Capital.*
20   Postone, *Time, Labor, and Social Domination*, 328.
21   Marx, *Capital*, Vol. 1, 443.
22   Ibid., 447.
23   Postone, *Time, Labor, and Social Domination*, 331.
24   Lewis Mumford, "Tool-Users vs. Homo Sapiens and the Megamachine," in Robert Scharff and Val Dusek (eds.), *Philosophy of Technology: The Technological Condition* (Malden, MA: Blackwell, 2003 [1967]).

Notwithstanding the transhistorical features that underpin the harnessing of socionatural powers that emerge from human cooperation, it is important to emphasize the historical specificity of the Marxian notion of the collective laborer. In the wake of the technological revolutions that followed the original introduction of automated systems of machinery in the eighteenth and nineteenth centuries, the assemblage of dead and living labors has reached a new synthesis. With the development of the sociotechnical basis of modern capitalist production, the abstract human nature that is the collective laborer assumes ever more sophisticated and novel configurations, exerting its command over a larger quantity of individual concrete labors.[25] Innovations in technologies for transport, haulage, and containerization in particular, combined with the possibilities enabled by the IT revolution, have enabled an unprecedented degree of integration among previously disconnected geographies of labor. As Deborah Cowen notes, these quantum leaps in industrial technologies have "stretched the factory" far beyond itself, directly and deliberately blurring the boundaries between transport and other forms of productive labor.[26]

In this "global social factory," which has been fully materialized by the geo-economic shift toward the Pacific Ocean discussed in the previous chapter, the construction of a global proletariat has, for the first time in history, become a concrete possibility. According to Starosta, this landmark shift has entailed the transformation of the modes of existence of the international working class, revolving around a fourfold qualitative differentiation. First, it has involved expanding the productive attributes of those wage-laborers performing the more complex parts of the labor process—all the intellectual and scientific tasks required by the automation of machinery.[27] As Mandel observed early in the onset of the third machine age,[28] the frantic pace of technological innovation characteristic of the third technological revolution was

---

25   In fact, one of the general assumptions in the debate over the Anthropocene is viewing the human species as a force of nature: Elmar Altvater, "The Capitalocene, or, Geoengineering against Capitalism's Planetary Boundaries," in Jason W. Moore (ed.), *Anthropocene or Capitalocene? Nature, History, and the Crisis of Capitalism* (Oakland, CA: PM Press, 2016).

26   Cowen, *Deadly Life of Logistics*, 104.

27   Starosta, "Revisiting the New International Division of Labour Thesis."

28   Mandel, *Late Capitalism*.

directly contingent upon the widespread proletarianization of intellectual labor and the transformation of science into an independent sphere of capital accumulation in its own right.

With the "university explosion," fostered by leaps and bounds in investment for research and development, Mandel noted, the main task of universities was no longer to produce educated people but rather to produce intellectually skilled wage-earners for the production and circulation of commodities.[29] The result of this was an overspecialization that translated into "expert idiocy" and entrenched intellectual divisions of labor, which, for Mandel, made intellectual workers unable to achieve an understanding of society as a whole.[30] The proliferation of research assistants, graduate students, and adjunct professors coping with three jobs in order to be able to cover their costs of reproduction, has come to embody Marx and Engels's claim that the bourgeoisie has stripped the halo from even the most honorable and respected professions.[31] Ursula Huws's notion of the *cybertariat* describes this emerging intellectual workforce, whose expanded productive attributes contrast with the precariousness of its living conditions.[32] Smith argues that existing theories on the real subsumption of nature tended to bypass this crucial element, because the subordination of the natural world to capitalist science and technology also involves reinventing *human biology* and harnessing the cooperative drives intrinsic to it.[33]

Besides purely intellectual work, Starosta shows that the commodification of scientific labor has also encompassed multiplying the human capacity to incorporate science in the immediate process of production.[34] This development will be illustrated in the next section, which discusses the evolution of the technical composition of labor in the mining industry. Second, the tendency to upskill the productive attributes of some organs of the collective laborer has gone hand in hand with the deskilling of others, because codified knowledge has become more directly incorporated into industrial technologies and hence machinery has emancipated itself even further from living labor. As Mumford

---

29   Ibid., 261.
30   Ibid., 265.
31   Marx and Engels, *Communist Manifesto*.
32   Huws, *Labour in the Global Digital Economy*.
33   Smith, *Nature as Accumulation Strategy*.
34   Starosta, "Revisiting the New International Division of Labour."

prophetically admonished, the afterlives of the megamachine in the age of modern technics were destroying the capacities for autonomous activities acquired by the human species over millennia, as the residual mass of workers were left to the trivial tasks of "watching buttons and dials, and responding to one-way communication and remote control."[35] A third tendency has, then, consisted in the extended production of processes of vertical reintegration, where nonmechanized tasks act as an external department of large-scale industry. As the hundreds of thousands of workers assembling iPads at Foxconn City in the Pearl River Delta paradigmatically exemplify, some industries have remained heavily dependent upon the manual skills of laborers.[36] Other industries, such as textiles and certain forms of resource extraction (coltan mining, rubber tapping), have been particularly resistant to mechanization given the impossibility of replacing the subtlety of the human hand's movements and other "natural" constraints of the labor process.[37]

Finally, a fourth layer in this pattern of differentiation entails the systematic production of surplus populations that are not formally employed but act as an "industrial reserve army." Usually composed of formerly free peasants, precarious migrants, and urban underclasses, the surplus population has traditionally functioned as a demographic mechanism to press down wages, to secure a constant supply of cheap labor-power (often commodified at a price below its cost of reproduction), and to discipline the workforce.[38] Because the relative surplus population is a byproduct of the concentration of social wealth in capitalist society, it is usually construed as an "externality" of the class struggle proper. However, because it has been systematically harnessed as an instrument to break worker solidarity and discipline workforces, Marx considers it an internal element of accumulation in its own right. For Marx,

---

35   Mumford, "Tool-Users vs. Homo Sapiens," 350.

36   Steven McKay, *Satanic Mills or Silicon Islands? The Politics of High-Tech Production in the Philippines* (Ithaca: Cornell University Press, 2006); Starosta, "Outsourcing of Manufacturing."

37   Starosta, "Revisiting the New International Division of Labour Thesis."

38   Deborah Cowen and Amy Siciliano, "Surplus Masculinities and Security," *Antipode* 43 , no. 5 (2011): 1516–41; Michael McIntyre, "Race, Surplus Population and the Marxist Theory of Imperialism," *Antipode* 43, no. 5 (2011): 1489–1515; Susanne Soederberg, *Debtfare States: Money, Discipline, and the Relative Surplus Population* (New York: Routledge, 2014).

if a surplus population of workers is a necessary product of accumulation or the development of wealth on a capitalist basis, this surplus population also becomes, conversely, the lever of capital accumulation, indeed it becomes the condition for the existence of the capitalist mode of production.[39]

As a result of these transformations in the materiality of the labor process, two conflicting tendencies account for the production and ongoing reproduction of the global collective laborer as a form of planetary nature or working organism. First, these forms of exertion of human labor-power do not exist in isolation but are premised upon unprecedented interdependence and socialization. It is therefore not possible to think about the degraded productive subjectivity of workers in a coltan mine or Foxconn assembly plant without intellectual wage-earners at other points of the supply chain codifying the knowledge to be incorporated in the former's instruments of production. Second, the tendency toward labor socialization unfolds alongside the internal polarization of the collective laborer in accordance with the different productive attributes that its members embody. According to Starosta, this growing differentiation has been at the basis of the formation of distinctive national and regional spaces of accumulation during the last four decades.[40] But these patterns of commodification of labor-power have also become projected into the built environment, rendering new constellations of land-use intensification, socioenvironmental change, and settlement space at local and regional scales.

Although planetary in scope, the socionatural entity that is the global collective laborer is therefore far from homogeneous. It is always contingent upon the production of internal differentiation on the basis of race, gender, ethnic-religious difference, and citizenship. As Chibber shows in his account of late industrialization in India, the expanded reproduction of capital is contingent upon racial and sociocultural differentiation, and sometimes even deliberately creates such diversification to ensure its ongoing reproduction.[41] This is one of the aspects where the

---

39   Marx, *Capital*, Vol. 1, 784.
40   Starosta, "Revisiting the New International Division of Labour Thesis."
41   Chibber, *Postcolonial Theory*; Anna Tsing, "Supply Chains and the Human Condition," *Rethinking Marxism* 21, no. 2 (2009): 148–76.

encroaching racial differentiation characteristic of old Western imperialisms has been rehashed and intensified in substantially more ruthless and pervasive ways than ever before. The production of a cultural "other," typical of the Cartesian mindset of Western colonialism, may have produced certain economic effects in previous stages of capitalist development. Such effects, however, existed in relations of exteriority with the core motion of capital accumulation. As Ferguson and McNally show,[42] the production of a hierarchically structured global market of human labor-power has linked geographies of labor in a complex social whole constituted by racialized forms of citizenship and nonmembership. When seen in these terms, these authors suggest, the social reproduction of the global working class entails processes of migration and racialization that are not incidental but inseparable from class and gender.[43] Thus Mezzadra and Neilson argue that, instead of being a collateral phenomenon, racialization lies at the very heart of the modalities through which bearers of labor-power become structurally produced as workers.[44]

The very existence of contemporary spaces of production such as *maquilas*, sweatshops in the textile industry, assembly plants for electronics components, and industrial household cleaning, are among the genuine products of this process of ethnoracial "othering" as internal to the production of the commodity labor-power. Present-day geographies of extraction, as we will see in subsequent sections, are likewise profoundly crisscrossed by ethnic, racial, and gender differentiation, as well as by the fragmentation and uneven development of the productive attributes of workers.

Spaces and Scales of the Mining Proletariat

The site of extraction performs a double alchemical transformation. It remotely feeds the life of distant cities by supplying raw materials that will later become remodeled in urban form, assuming the social form of commodities. Yet spaces of extraction also constitute a locus of

---

42 Ferguson and McNally, "Precarious Migrants."
43 Ibid., 3.
44 Mezzadra and Neilson, *Border as Method*, 20.

urbanization in themselves because they exert a definitive imprint upon the geography as a result of transfers of technology, flows of investment, shifting frameworks of interaction, and reconfiguration of the built environment. Ever since the dawn of modern technics, when the methods and ideals of mining became the chief pattern for industrial effort across the Western world, the whole animus and mindset of mining have expanded to revolutionize the entire economic and social organism in the rural lifeworld.[45] In addition to its radical effects on the built environment, Mumford noted how the recklessly instrumental, "get-rich-quick" attitude of mining was systematically dismantling rural and premodern life and culture. In a similar vein, van der Pijl explains that the common characteristic of the imposition of the discipline of capital as it colonizes its constitutive outside resides in the breaking of the traditional mold of agrarian existence by commodification. This process, van der Pijl suggests, is most starkly manifested in the urbanization of the countryside.[46]

In contemporary spaces of extraction, the extension of the commodity relation—and hence of the spatiotemporality of capital—to rural life has assumed novel and ever more pervasive configurations. The evolution of the social and technical composition of labor in Latin America's mining industry reflects the qualitative technological shifts that are intrinsic to the present configuration of the capitalist mode of production generally considered. Perhaps one of the most striking trends in the mining industry is its proclivity to mirror regimes of industrial organization implemented in the electronics industry, especially concerning the increasing relevance and participation of large transnational contractors in the core business. Increasing capitalization and technological sophistication triggered by declining mineral grades, coupled with financially driven managerial strategies to streamline and reduce operation costs systematically, has made the mining supply chain resemble the "modular" and "turnkey" production networks described in the previous chapter.[47] Roughly put, this means that the mining industry has moved toward a combination of operational flexibility with vertical

---

45   Mumford, *Technics and Civilization*.

46   Pijl, "International Relations and Capitalist Discipline."

47   Timothy Sturgeon, "Modular Production Networks: A New American Model of Industrial Organization," *Industrial and Corporate Change* 11, no. 3 (2002): 451–96; Starosta, "Outsourcing of Manufacturing."

reintegration at the top of the supply chain. Although subcontracting has traditionally figured as a key component of mining activity, recent years have marked a shift toward greater reliance on large-scale contractors. Because of this novel organizational configuration, mining companies have been able to commodify labor-powers of heterogeneous complexity under a systemic industrial vision.

In 2004, the copper-mining industry in Chile advanced $5 billion in operational expenditures, of which $1.3 billion went to acquiring goods and inputs, and $1.7 billion to nonstrategic services.[48] As industrialization in East Asia advanced in tandem with international demand for raw materials, mining companies became increasingly reliant on contractors. By 2013, mining corporations were allocating 60 percent of their operational costs for the acquisition of goods and services from third parties.[49] Aside from being a relatively new sector in the Chilean economy, the contracting industry for mining has also become very dynamic. In 2010, there were 4,643 registered contractors, while in 2012 the figure rose to 5,998, which means that in only two years the sector grew by 29 percent.[50] Besides its dynamism, the sheer size of the sector attests to the increasing relevance of contractors for the mining industry as a whole. In 2012, the sector accounted for 7.4 percent of Chile's GDP, while the mining industry as a whole accounted for 12 percent.[51] As of 2012, contractors employed 712,697 laborers under direct working contract, equivalent to 10 percent of Chile's working population.[52]

The tasks initially outsourced in the mining industry were labor-intensive and marginal to the core operation, such as catering, hostels, and shaft sinking, but gradually began to include more complex services, such as mineral forecasting, geological modeling, and engineering. The trend toward a larger degree of diversification in services provided seems to have taken off in the South African mining industry

---

48   COCHILCO, *Oportunidades de Negocios para Proveedores de Bienes, Insumos y Servicios Mineros en Chile* (Santiago: COCHILCO, 2005), 268, figure 5.

49   Innovum/Fundación Chile, *Proveedores de la minería chilena: Estudio de caracterización 2014* (Santiago: Fundación Chile, 2014), 2.

50   Ibid.

51   Ibid., 2, 11.

52   Fundación Chile, *Proveedores de la minería chilena: Estudio de caracterización 2012* (Santiago: Fundación Chile, 2012), 18.

during the 1990s, when whole shafts began to be outsourced to third parties.[53] In Chile, this trend has evolved to the extent that large transnational corporations now figure among the key actors in the contracting sector for mining. Komatsu, Siemens, and Finning Cat are among the companies that commonly operate alongside extractive corporations in shafts, pits, and industrial facilities for mineral processing.[54] The distribution of the services outsourced to contractors illustrates the evolving technical composition of labor in the extractive industries. Labor-intensive tasks such as transport, catering, and security constitute the largest percentage of services provided. The more complex, capital-intensive tasks, such as construction, mechanical engineering, industrial and electrical equipment, drilling, mineral crushing, explosives, and lab analysis, account for only 9 percent of outsourced services (see figure 5).

Figure 5 Distribution of services provided to the mining industry in Chile

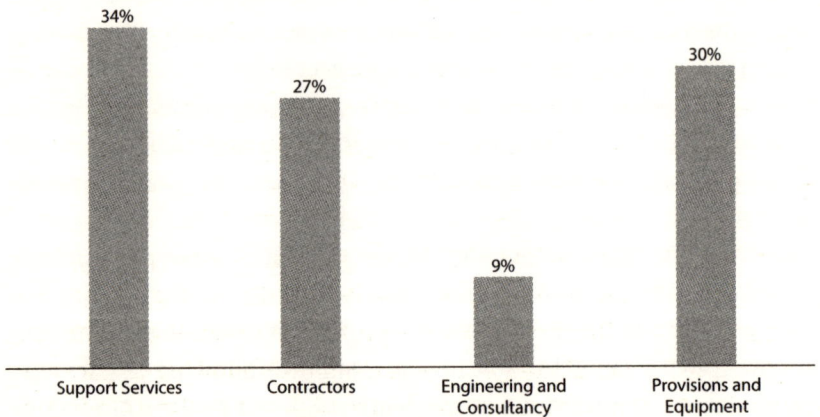

Source: *Innovum/Fundación Chile*, Proveedores, 2014, 8.

These figures reveal a trend that is not particular to the mining industry but symptomatic of an overall shift in global labor organization,

53   Bridget Kenny and Andries Bezuidenhout, "Contracting, Complexity and Control: An Overview of the Changing Nature of Subcontracting in the South African Mining Industry," *Journal of the South African Institute of Mining and Metallurgy* (July/ August 1999), 185–92.
54   Innovum/Fundación Chile, *Proveedores, 2014.*

where the tendency toward automation unfolds alongside internal polarization between the various organs of the collective laborer in accordance with their productive attributes. High-skill services such as engineering, consultancy, and lab analysis are contingent upon the expansion of the productive capacities of a segment of the workforce, so the core operations of the mine tend to be carried out by a clique of well-remunerated and skilled wage-laborers. As mines tend to be geographically remote from cities, this "privileged" organ of the collective laborer is, almost without exception, composed of city dwellers with specialized academic training and no connection with the sites of extraction. As an official of Codelco remarked,

> The new generations of engineers and technicians do not want to spend 12 hours a day stuck in the dirty, gruesome environment of a mining shaft or pit. They are "technological by nature," and so they prefer to operate the mine from remote locations, using joysticks and digital platforms.[55]

The more specialized the worker, the farther away she generally lives from the extraction site. The most qualified engineers usually live in Santiago or other Latin American capitals and commute to extraction sites when required. When these engineers visit the sites, Merchant notes, they stay in "lavish base camps" akin to "a tiny five-star hotel with ten or so rooms and a private chef plunked down in the weird alien desert."[56] The wealthy and extravagant lifestyles of mining engineers are perhaps nothing particularly specific to the twenty-first century. As Gray Brechin shows, the profession of mining engineering acquired a heroic, almost evangelical stature following the wave of technological innovation and mechanization tied to the introduction of the internal-combustion engine in the late nineteenth century.[57] The mining engineer of earlier phases of mechanization, according to Brechin, was "tough yet refined, moving as easily amid the smelters of South Africa's Rand and Butte . . . as in the palaces and bourses of the

---

55   Interview with a Chilean mining executive, January 10, 2017.
56   Merchant, *One Device*, 114.
57   Gray Brechin, *Imperial San Francisco: Urban Power, Earthly Ruin* (Berkeley: University of California Press, 2006).

European capitals or the New York Stock Exchange."[58] For an ambitious young man in that day and age, Brechin considers, "a mining career offered the chance to win the wealth capable of propelling him into the same Olympian caste as the men and women for whom he worked."[59]

Santiago de Chile, in particular, has become one of the most attractive cities in Latin America for the highly skilled workers, financiers, and executive personnel of the mining industry. Many transnational firms and contractors have settled in this bustling, modern city. According to De Mattos, the factors that have made Santiago an attractive hub for transnational corporate networks are its communications/transport infrastructures, its physical proximity to other firms, the availability of a wide array of production services, its highly diverse and qualified labor markets, and relatively diversified industrial base.[60] As a result of the urban modernization that followed Chile's neoliberal transformations after the 1970s, De Mattos explains, Santiago became increasingly positioned as an organizational hub for global economic operations, similar to other "global cities."[61] According to Riffo Pérez, Santiago is where the firms and economic groups that plan, coordinate, and control Chile's export-oriented sectors have established their headquarters.[62] Perhaps mirroring the growing outsourcing of services and tasks in the mining industry, the service sector that has displayed the fastest growth rates in Santiago. In general terms, it has been argued that Santiago's introduction to transnational corporate networks has upgraded the technical composition of the local workforce.[63] Not surprisingly, then, 62 percent of the mining industry contractors are based in Santiago.[64]

---

58   Ibid., 53.

59   Ibid.

60   Carlos de Mattos, "Santiago de Chile: metamorfosis bajo un nuevo impulso de modernización capitalista," in Carlos de Mattos, María Elena Ducci, Alfredo Rodríguez, and Gloria Yáñez Warner (eds.), *Santiago en la globalización: una nueva ciudad?* (Santiago: Ediciones SUR, 2005).

61   Ibid.

62   Luis Riffo Pérez, "Los impactos de la globalización sobre los mercados de trabajo metropolitanos: el caso de Santiago de Chile en la década de los noventa," in De Mattos et al. (eds.), *Santiago en la globalización.*

63   Ibid., figure 6.

64   Innovum/Fundación Chile, *Proveedores, 2014,* 10.

Figure 6 Employment growth in Santiago according to occupational group, 1992–2002

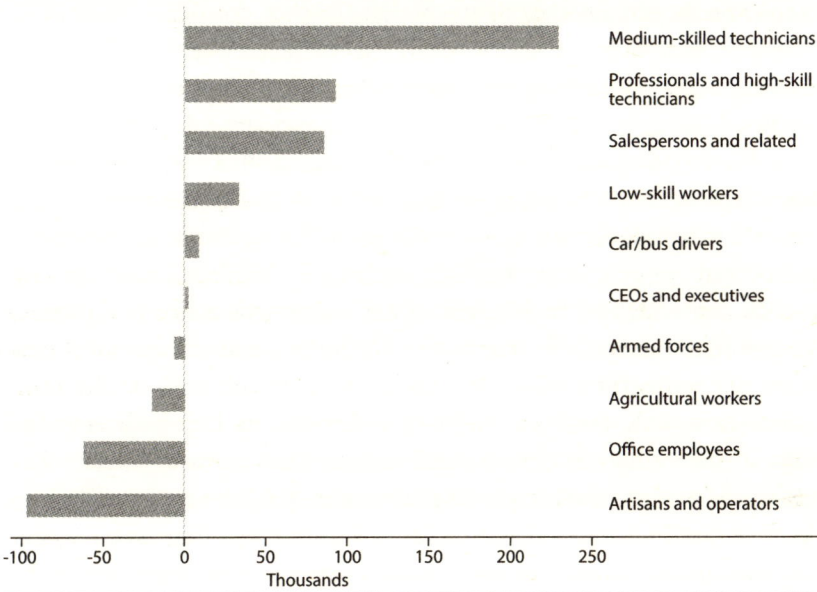

Medium-skilled technicians

Professionals and high-skill technicians

Salespersons and related

Low-skill workers

Car/bus drivers

CEOs and executives

Armed forces

Agricultural workers

Office employees

Artisans and operators

-100   -50   0   50   100   150   200   250
Thousands

Source: Riff o Pérez, "Los impactos de la globalización," 178.

For its own self-preservation, Mandel explains, capital could never afford to transform all workers into scientists, just as it could not transform all production into automation.[65] Thus, concomitant with the expanded forms of productive subjectivity required to operate highly sophisticated mechanized systems of extraction and mineral processing, the rest of the mining supply chain revolves around degrading types of labor-power. As in the electronics industry,[66] this social process has superimposed the formal mediations of citizenship, race, gender, and ethnicity upon the material differentiations of fractions of the mining proletariat. The degraded forms of labor-power that either act as appendages of industrial systems or perform manual tasks such as cleaning or cooking constitute the mass of the workforce. Their territorial distribution often conflicts with their social composition, as the middle tier of this workforce comes from other places across Chile and sometimes even from neighboring countries. Since mining and energy projects tend to require large amounts of unskilled industrial labor of this sort to work temporary "seven by

---

65   Mandel, *Late Capitalism*, 208.
66   Starosta, "Outsourcing of Manufacturing."

seven" shifts (seven days on site, seven days away), mining towns have received large numbers of floating populations.[67]

These temporary workers—informally called *faeneros* by local communities—come from various locations on a temporary basis, have no attachments to the host town, are usually underpaid, and face over-crowded accommodations. As a result, social ills that were uncommon before the commodity boom, such as sex work, theft, street fights, drug abuse, and sexual assaults, are now common.[68] Because these towns tend to be intensive on energy, logistics, and mining infrastructure, they are usually burdened by high levels of air, water, and noise pollution, a feature that dramatically impacts public health and the quality of life more generally. Informally referred to in Chile as *zonas de sacrificio* (sacrifice zones), these are the built environments that have come to support the modalities of commodification of labor-power correspond-ing to the organs of the collective laborer that act as the appendage of heavy infrastructures and systems of machinery. Practices of outsourc-ing and subcontracting are predominant among these segments of the workforce, increasing dramatically during recent years (see figure 7). In the mining industry, as in many other industries, temporary contracts have become a breeding ground for deskilling and income inequality. As Cademartori illustrates, in one of the largest mining companies operat-ing in Chile, workers directly employed by the firm earn on average 254 percent more than those hired by external and local contractors.[69]

Finally, acting as the lower tier in the mining workforce is a further organ of the collective laborer in charge of nonmechanized/manual tasks, which can be considered to constitute an "external department" of large-scale industry proper, as discussed in the previous section. This lower rank encompasses a wide array of services that include formal activities such as cleaning, catering, and security, as well as less formal ones such as street vending, payday lending, tourism, and sex work. As Charmaine Chua observes, the sophisticated sociotechnical basis of

---

67 Martín Arboleda, "In the Nature of the Non-City: Expanded Infrastructural Networks and the Political Ecology of Planetary Urbanisation," *Antipode* 48, no. 2 (2016): 233–51.

68 Ibid.

69 Jan José Cademartori, *Inversión extranjera en el desarrollo de la región minera de Antofagasta: historia y perspectivas* (Antofagasta, Chile: Universidad Católica del Norte, 2010), 259.

Figure 7 Share of subcontracted workers in Chile's mining industry, 1975–2004

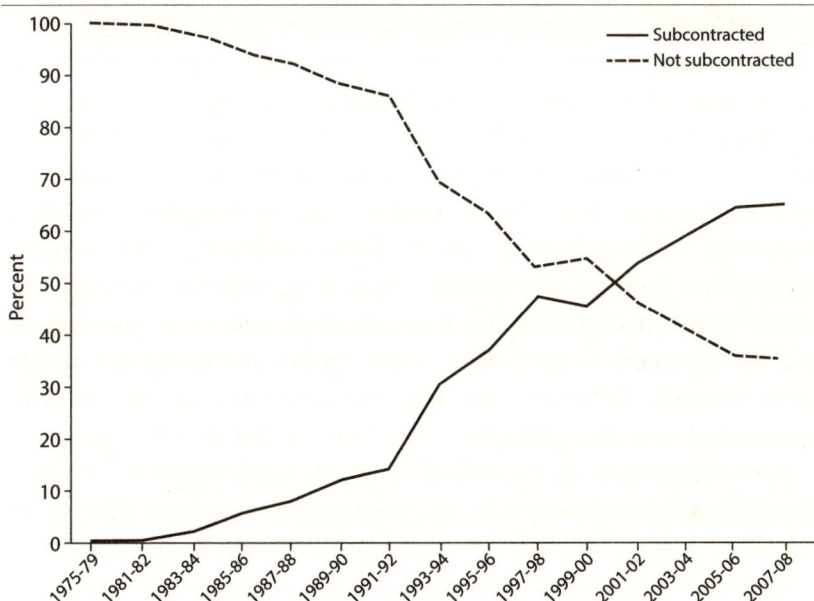

Source: Arias et al., "Large mining enterprises," 86; Miguel Atienza, Marcelo Lufin, and Juan Soto, "Mining Linkages in the Chilean Copper Supply Network and Regional Economic Development," Resources Policy, February 13, 2018.

contemporary logistics is riddled by contradictions similar to those of the mining industry.[70] Lurking behind the homogenizing fantasies of the kinetic, clockwork mechanical apparatuses of Chinese ports, she points out, there is a shadow economy of precarious laborers tasked with pumping dirty fuel out of tanks, collecting accumulated garbage and scrap, and cleaning used pipes.

In the Chilean mining industry, the degraded organs of the collective laborer are internally polarized and fragmented in terms of the productive attributes that capital demands from each category of labor-power. They are also deeply differentiated according to gender and ethnoracial attributes. In general, and because Chile has developed relatively homogeneous processes of *mestizaje* when compared to other Latin American countries (such as Colombia, Venezuela, and Brazil), the question of race has only recently aroused interest from academics and the wider public. Although there are no comprehensive studies addressing the racialization of work in the mining industry of Chile or Latin America

70   Charmaine Chua, "The Chinese Logistical Sublime and Its Wasted Remains," *The Disorder of Things*, 2015, thedisorderofthings.com.

in general, evidence from South Africa suggests that mining corporations have tended to systematically introduce *de facto* "color codes" for different categories of labor-power.[71] This, it is argued, constitutes a managerial strategy to press down wages, fragment the workforce, and extend the discipline of capital over labor.[72] As Lüthje et al. illustrate,[73] ethnoracial and cultural attributes are also a central element in segmenting workforces across the electronics-manufacturing industry. According to these authors, large manufacturing contractors tend to hire workers who are migrants and racial minorities for low wage and insecure jobs. In most low-cost locations, they point out, "the employment of a workforce with these characteristics is possible because of complex systems of labor migration and ethnic, religious, and cultural division and discrimination."[74]

Seen in this light, the patterns of labor differentiation taking place in the mining industry resemble those being systematically implemented across the large global contractors of the electronics industry. In most cases, the difference in pay and working conditions between manufacturing workers and engineers, technical and administrative personnel, can be gigantic.[75] In the words of Lüthje and coauthors, this "reflects the companies' efforts to attract engineering and technical talent while squeezing labor costs in manufacturing."[76] In the mining industry, this uneven distribution of productive capacities and attributes is not engendered in a vacuum but forged through diverse mechanisms and strategies of violence, enclosure, racialization, labor casualization, and displacement. In this sense, the mining proletariat is a microcosm of a process that encompasses the whole globe and consists in the systematic incorporation of erstwhile self-subsisting peasantries and other agrarian and indigenous communities into the polarizing industrial organism that is the global collective laborer. The next section interrogates the latter process.

---

71   Kenny and Bezuidenhout, "Contracting, Complexity and Control."
72   Ibid.
73   Lüthje et al., *From Silicon Valley to Shenzhen*.
74   Ibid., 155.
75   Ibid.
76   Ibid., 227–8.

Primary-Commodity Production and the Twilight of the Peasantry

As previously noted, the universalization of the commodity form is contingent upon greater socialization of labor and greater interdependence between individuals. When the social formation based on this form of interdependence becomes fully developed, Postone stresses, it acquires a necessarily systematic character and supersedes other social forms, becoming global in scale.[77] The ongoing dismantling of those "other social forms" required for the material assembly of the global collective laborer is what agrarian historians and sociologists have termed *global depeasantization*,[78] or what Eric Hobsbawm evocatively refers to as the "twilight of the peasantry."[79] For Hobsbawm, the most fundamental change begotten by the twentieth century—even compared with world-historical events such as two world wars, decolonization, the IT revolution, and the rise and fall of state socialism—was undoubtedly "the passing of peasant and rural life."[80]

Superseding externalist conceptions of nature in studies of urbanization demands renewed attention to these transformations in humans' metabolic interaction with their "inorganic body"—the material nature of sensuous existence. According to Schmid, besides mere concentration of built environments and infrastructures, the urban is also a specific form of spatial praxis, woven together by networks of interaction in daily life.[81] For this reason, the process of planetary urbanization also encompasses the ongoing expansion of such networks of interaction (production, consumption, leisure) beyond densely populated areas. The concept of global depeasantization thus powerfully captures

---

77 Postone, *Time, Labor, and Social Domination*.

78 Araghi, "Global Depeasantization"; Araghi, "Accumulation by Displacement"; McMichael, "Peasant Prospects in the Neoliberal Age"; Vanhaute, "Peasants, Peasantries, and (De)peasantization"; Kay, "La transformación neoliberal."

79 Eric Hobsbawm, *The Age of Extremes: The Short Twentieth Century 1914–1991* (London: Abacus, 1994), 289.

80 Cited in Farshad Araghi, "The Great Global Enclosure of Our Times: Peasants and the Agrarian Question at the End of the Twentieth Century," in Fred Magdoff, John Bellamy Foster, and Frederick Buttel (eds.), *Hungry for Profit: The Agribusiness Threat to Farmers, Food and the Environment* (New York: Monthly Review Press, 2000), 158.

81 Christian Schmid, "Networks, Borders, Differences: Towards a Theory of the Urban," in Brenner (ed.), *Implosions/Explosions*.

the expansion of these forms of urban spatial praxis as it expresses the generalized assault on rural populations that followed the industrialization of the global South. Araghi notes how, at the onset of neoliberal capitalism, Third World peasantries with direct ownership of the means of subsistence became systematically exposed to world market forces via the price form.[82] This entailed an unprecedented penetration of commodity relations into the countryside, forcing peasants to diversify their sources of income (by selling their labor-power to landlords or multinational corporations) or even to migrate to cities.

The ongoing, massive dispossession of world peasantries, Araghi claims, has allowed capital to appropriate vast quantities of migratory surplus labor-power as well as to accumulate spaces of surplus nature—a process often referred to as "planetary land grabs."[83] This world-historic shift in the social composition of the international working class has been predicated upon the dismantling of agricultural subsidies, the mechanization of primary-commodity production, the reconstruction of the value relations that had been undermined by state developmentalism, and the redistribution of publicly funded subsidies to the advantage of centralized, agroindustrial capitals.[84] Economic liberalization has, then, consisted in relentlessly displacing previously self-reproducing peasantries into an expanding circuit of casual labor, flexibly employed when employed at all.[85] The material articulation of these forms of degraded productive subjectivity has contributed to the dramatic, metastasizing growth of shantytowns across the cities of the so-called global South, profoundly eroded agrarian ways of living, and provided an endless supply of cheap labor-power for capital to consume, sometimes even at below its cost of reproduction.

The global industrial restructurings of the fourth machine age, which have hinged upon the relentless and unprecedented explosion of manufacturing enclaves, special economic zones, maquilas, agroindustrial hinterlands, and open-cast mines, would have been unthinkable without the generalized dispossession of world peasantries. To begin with, industrial upgrading in postsocialist China would have never come to

---

82   Araghi, "Global Depeasantization."
83   Araghi, "Accumulation by Displacement."
84   Ibid.
85   McMichael, "Peasant Prospects in the Neoliberal Age."

fruition without the hundreds of millions of peasants being severed from their environs and transformed into degraded organs of the urban-industrial workforce. But there are additional, equally stark examples of this trend. Arundhati Roy shows that the Delhi Mumbai Industrial Corridor, a projected 1,500-kilometer tract of land with nine industrial megazones, is set to conjure such vast forces of production that its aggregate population will go from 214 million in 2014 to 314 million by 2019.[86] Such vast movements of people are far from trivial, and in the case of India have required concerted actions from technocrats, ruling elites, landed aristocracies, the military, and even death squads. Amid encroaching industrialization, Roy reports, around 250,000 farmers have died by suicide to flee the horrors of punishing debt, harsh proletarianization, and even hunger.

Although the figures are not as bewildering in Latin America as in Asia, the phenomenon is essentially the same. During the last decade, an estimated 900,000 peasants in Paraguay have been displaced by landlords and state forces in order to enable the expansion of capital-intensive monocultures.[87] In Argentina, an estimated 200,000 rural families have been displaced by transgenic soybean production.[88] Due to long-standing armed conflict, Colombia provides perhaps the most striking case of depeasantization in the region. An estimated 6 million peasants have been displaced by agroindustrial elites and paramilitary groups in Colombia during recent decades to make way for mining, energy, and agroindustrial investments, or just land speculation and money laundering.[89]

Because Latin American economies have become specialized in producing primary commodities, depeasantization in the region is

---

86   Arundhati Roy, *Capitalism: A Ghost Story* (Chicago: Haymarket, 2014), 17.

87   Santi Carneri, "La codicia por la tierra en Paraguay," *El País*, March 2, 2017, elpais.com.

88   Ingrid Feeney-McCandless, "Por una Vida Digna: Science as Technique of Power and Mode of Resistance in Argentina," *Alternautas: (Re)Searching Development*, September 22, 2017, alternautas.net.

89   Forrest Hylton, *Evil Hour in Colombia* (New York: Verso, 2006); Teo Ballvé, "Everyday State Formation: Territory, Decentralization and the Narco Landgrab in Colombia," *Environment and Planning D: Society and Space* 30 (2012): 603–22; Nubia Yaneth Ruiz Ruiz and Luis Daniel Santana Rivas, "La nueva geografía de la explotación minero-energética y la acumulación por desposesión en Colombia entre 1997 y 2012," *Notas de Población* 102 (January/February 2016): 249–77.

directly connected with the expansion and reproduction of natural-resource frontiers. According to Ruiz Ruiz and Santana Rivas, the mining concessions granted by the Colombian state since the late 1990s are directly connected with the violent expulsion of peasants from the land.[90] The regions where mining and energy megaprojects have tended to expand more forcefully, these authors argue, are precisely those where forced displacement and systematic intimidation of peasants have taken place more blatantly. As an outcome of an inherently racialized and geographically uneven process, this mass of dispossessed humanity is overwhelmingly indigenous and Afro-Colombian.

In general terms, selective killings, death threats, and constant intimidation of peasant and indigenous leaders have become a common feature of mining and agroindustrial projects in many Latin American countries. In 2015, fifty *campesino* leaders were murdered in Brazil, followed by twenty-six in Colombia, twelve in Peru, ten in Guatemala, eight in Honduras, and four in Mexico.[91] Fleeing violence, environmental destruction, or starvation, these groups have been forced to migrate to the shantytowns of cities or to other parts of Latin America, which recently have come to include the mining districts of northern Chile. A considerable part of these displaced peasants, who come from Colombia and across Latin America, have thus become subordinated to the technical division of labor of primary-commodity production as its industrial reserve army, dramatically swelling the shantytowns of cities such as Antofagasta and Calama, a mining town in northern Chile. It is estimated that the number of families living in *campamentos* in Antofagasta skyrocketed from 632 in 2007, to 6,229 in 2016.[92] Dark-skinned migrants from Bolivia and Peru are usually hired as gardeners, technicians, and subcontracted *faeneros* for the mining industry; Afro-Colombian migrants tend to work in the informal economies of mining towns as street vendors, plumbers, and construction workers.

Gender is also a constitutive element in the types of forced displacement and precarious migratory trends that emerge from primary-commodity production. The expansion of resource frontiers has led

---

90   Ruiz Ruiz and Santana Rivas, "La nueva geografía."

91   Greenpeace, "Brazil: The Most Dangerous Country for Environmental Activists," Greenpeace International, 2016, greenpeace.org.

92   Techo Para Chile, "Catastro de Campamentos 2016," 60.

women, in some cases, to lose their husbands due to armed violence and then be intimidated and threatened to sell their plots to mining companies.[93] This phenomenon, referred to as *madresolterismo*, has become a systematic trend in the Colombian countryside. According to a census by the United Nations, 81.6 percent of women in Colombia's rural areas are single heads of family, and this puts them in harsh conditions of social vulnerability and often also extreme poverty.[94] Many are faced with no option but to migrate, and this explains to a certain extent the feminization of the migrant population in the mining towns of Chile.[95] Migrant women usually develop informal activities such as fortune telling and manicure and beauty-parlor services, and often fall prey to human-trafficking networks, being forced to work as prostitutes.[96] They tend to be vulnerable to racial discrimination and sexual violence, as they are usually hypersexualized, bearing the burden of crude racial stereotyping, especially in the midst of the macho culture that predominates in mining towns.[97]

For Iñigo Carrera,[98] processes of racialization that are triggered by migration tend to reproduce national boundaries within state territory through differentiation on the basis of citizenship or nonmembership. Movements of surplus populations across national borders, Ferguson and McNally point out, accrue a twofold advantage for capital because they fragment the national working class and press down the cost of

93   Rosa Emilia Bermúdez Rico, "Impactos de los grandes proyectos mineros en Colombia sobre la vida de las mujeres," in Toro Pérez, Fierro Morales, Coronado Delgado, and Roa Avendaño (eds.), *Minería, Territorio y Conflicto en Colombia* (Bogotá: Universidad Nacional de Colombia, 2012).

94   Bermúdez Rico, "Impactos de los grandes proyectos mineros"; Astrid Ulloa, "Feminismos territoriales en América Latina: defensas de la vida frente a los extractivismos," *Nómadas* 45 (October 2016): 123–39.

95   Echeverri, "Otredad racializada."

96   Connectas, "El Nuevo éxodo latino: De Colombia a Chile," *Connectas/El Mercurio*, 2016, connectas.org; Mónica Amador Jiménez, "La incesante diáspora Africana: afrocolombianas solicitantes de asilo en el norte chileno," *Nomadías* (2011), 89–103; Echeverri, "Otredad racializada."

97   Amador Jiménez, "La incesante diáspora Africana"; Liberona Concha, "La frontera cedazo"; María Emilia Tijoux, "El Otro inmigrante negro y el Nosotros chileno: un lazo cotidiando lleno de significaciones," *Boletín Onteaiken* 17 (May 2015): 1–15; Thodes Miranda, "Segregación socioespacial en ciudades mineras."

98   Iñigo Carrera, *El Capital*.

labor-power.[99] They also effect a rescaling of the household on the basis of a transnational separation between production and reproduction. Migrants often leave their families behind and send remittances back to their home countries, dramatically reducing the cost of reproducing current and future generations of workers.[100] In Antofagasta, Echeverri shows that women spearhead migratory processes and tend to work two shifts a day to support themselves and their families.[101] This allows them to cover their costs of reproduction and also send some money to their children while they implement diverse strategies to reunite with them. In the words of a migrant woman living in Antofagasta, "Many women leave their children there [in Colombia], because with the two shifts one needs to do in order to get by and also send money home, it is impossible to take care of them."[102]

The spatial imprint of wide-ranging global depeasantization is most patently manifested in the staggering, explosive growth of precarious and "informal" settlements in recent decades. Araghi, for example, estimates that 65 percent of the growth in urban populations in the 1980s and 1990s was attributable to rural-urban migrations.[103] This phenomenon, which accelerated sharply during the neoliberal period, reached its pinnacle in the aftermath of the financial meltdown of 2008. Teresa Caldeira conceptualizes these modalities of spatialization as "peripheral urbanization," a type of city-building that is always makeshift, irregular, and renders highly unequal and geographically uneven urban environments.[104] These geographies of indentured and precarious labor push us to consider the close connection between the modern wage system and extraction. Writing about the trans-Atlantic slave trade, Mbembe describes how African people were transformed into "living minerals" from where wealth was extracted by forcible means. This

---

99   Ferguson and McNally, "Precarious Migrants."
100   Ibid.; Nicola Phillips, "Migration as Development Strategy? The New Political Economy of Dispossession and Inequality in the Americas," *Review of International Political Economy* 16, no. 2 (2002): 231–59.
101   Echeverri, "Otredad racializada."
102   Ibid., 95.
103   Araghi, "Great Global Enclosure," 151.
104   Teresa Caldeira, "Peripheral Urbanization: Autoconstruction, Transversal Logics and Politics in Cities of the Global South," *Environment and Planning D: Society and Space* 35 , no. 1 (2016): 3–20; James Holston and Teresa Caldeira, "Urban Peripheries and the Invention of Citizenship," *Harvard Design Magazine* 28 (2008): 19–23.

gave rise to a transition from "*homme-minerai* to *homme-métal* to *homme-monnaie.*"[105]

In contemporary Latin America, Ulloa describes a similar process whereby violence is increasingly employed against men and women as they are rendered "bodies-territories" of appropriation and dispossession.[106] The encroaching "masculinization of space" that is concomitant with mining towns in Latin America, Ulloa highlights, reduces women to indentured laborers who often work under conditions of low pay and constant sexual harassment and abuse. The realization of a new period of technological and industrial development, then, by no means entails the erosion or drifting away of the restless movement of contradiction constituted by class relations. As Marxian interpretations of colonialism and slavery have recently shown, modern capitalist relations based on the commodification of labor-power are not anathema or even an iteration of a more developed form of the bondage and direct coercion that defined chattel slavery.[107] The two coexist as an amalgamated, organic whole.

As McNally shows, the process of abstraction that allowed the consolidation of the modern money form was facilitated by the slave trade.[108] To be enslaved, McNally contends, is to be abstracted from humanity; hence slavery bears an elementary logic of social abstraction and commensuration between people and things that foreshadows the totalizing logic of capitalist commodification. The commensuration that underpinned the development of the modern monetary system introduced by the Bank of England in the eighteenth century, according to McNally, was closely bound up with the pricing of slaves, particularly women in bondage.[109] Racial and gender domination should therefore not be understood as an expression of the territorial logic of state power, while labor exploitation is seen as the domain of capital accumulation. The two of them are already imbricated into the inner fabric of the real

---

105   Mezzadra and Neilson, "On the Multiple Frontiers of Extraction," 7.

106   Ulloa, "Feminismos territoriales en América Latina," 126.

107   McNally, "The Blood of the Commonwealth"; Nikhil Pal Singh, "On Race, Violence, and So-Called Primitive Accumulation," *Social Text* 128, no. 34–3 (2016): 27–50; Mbembe, *Crítica de la razón negra*.

108   McNally, "The Blood of the Commonwealth."

109   David Graeber, *Debt: The First 5,000 Years* (New York: Melville House, 2014 [2011]); Moore, *Capitalism in the Web of Life*, 213.

abstraction to such an extent that it does not make sense to consider them separate.

Aside from the spectacular violence of bulldozers, police trucks, bullets, and evictions, the depeasantization associated with resource extraction unfolds in tandem with slower temporalities, some of which are barely noticeable within months or even years. The history of the mining industry in Latin America offers a relevant vantage point from which to visualize how the urbanization process is slowly yet deeply inscribed in the subjective composition of the rural laborer. In Chile, Vergara Marshall recounts how mining companies introduced mechanized systems for energy production, refining, drilling, and crushing in the 1960s alongside Taylorist principles of scientific management.[110] These technical innovations were introduced not only to enhance the productivity of dead labor but to break obdurate rural habits and attitudes of living labor. Behaviors such as absenteeism, instability, and lack of a proper "work ethic," typically associated with semiproletarian culture, were targeted by corporate management as key obstacles to productivity.

Álvaro García Linera's historiographical account of the evolution of the mining proletariat in Bolivia since the late nineteenth century illustrates how the social and organizational basis of social production is dialectically interwoven with the erosion of rural ways of living.[111] He notes how early iterations of mechanized metal-ore extraction were associated with the productive subjectivity of the *artisan-operator*, who exerted her autonomous productivity within the industrial system of the mine.[112] Artisan-operators maintained strong ties to the communal-peasant productive structure, evident in their forms of resistance (the riot and the party) as well as in their use of time and the *cajcheo*.[113] In this sociotechnical configuration, living labor was only formally subsumed to capital, and human productive subjectivity was moored in agricultural or ecological temporality. During the phase of monopoly capital, García Linera illustrates how this form of

---

110   Ángela Vergara Marshall, "Conflicto y modernización en la gran minería del cobre (1950–1970)," *Historia* 37 , no. 2 (2004): 419–36.

111   Álvaro García Linera, *Plebeian Power: Collective Action and Indigenous, Working-Class, and Popular Identities in Bolivia* (Chicago: Haymarket Books, 2015).

112   Ibid.

113   *Cajcheo* is a peasant/indigenous worker practice of extracting and gathering minerals without any form of external control or supervision.

productive subjectivity eroded with the appearance of the skilled worker in large mining companies, whose embodied technical skill retained certain elements of its predecessor but was now situated within the shift from formal to real subsumption of labor.[114] This new sociotechnical reality led to a breakdown of the workers' strong ties with the agricultural world and to the emergence of what García Linera refers to as a properly "industrial rationality." Premodern understandings of social time, dietary customs, lifestyles, and work ethics were progressively superseded by behaviors typical of the industrial mindset, such as labor discipline, workplace association, the patriarchal family unit, and the commodification of the conditions of social reproduction.[115]

In the geographies of the planetary mine, depeasantization has accelerated substantially as a result of new technical configurations of social production, as well as of the transformative effect of information technologies in patterns of socialization. Smallholding and artisanal agricultural production have been gradually disappearing as a result of air, water, and noise pollution, which exerts devastating effects upon the integrity of soils and crops. Also, the arrival of mining projects to rural areas forms a breeding ground for income inequality and sociospatial exclusion, as wages for *faeneros* are substantially higher than those earned by smallholders and agricultural workers. As a result, many leave their jobs in agriculture to work for mining companies or their contractors. Younger generations aspire to someday join this temporary industrial workforce and afford urban lifestyles based on fast food, fast fashion, and smartphones. As subsequent chapters will illustrate in detail, retail finance has also revolutionized the social reproduction of local communities in the rapidly urbanizing countrysides of the commodity supercycle. Mining towns in northern Chile are littered with stores offering all sorts of clothing, electronic devices, and services. The computerization of the credit system has enabled financial actors to expand debt (in the form of credit cards, consumer loans, and even payday loans) to the everyday lives of rural populations experiencing abrupt sociospatial change.

114   García Linera, *Plebeian Power*, 109.
115   Ibid., 111.

Conclusion

As the case of the mining industry illustrates, the collective laborer assembled by the current industrial era has spearheaded a modality of human integration and interdependence that is without precedent. The megamachine, which Mumford describes to illustrate the cooperative powers that made possible ancient marvels of engineering in Egypt and Mesopotamia, appears as insular, petty, and insignificant in the face of the geographies of labor that have emerged under the aegis of supply-chain capitalism. Moreover, contemporary spaces of extraction provide a relevant vantage point from which to grasp the rampant sociospatial and ethnoracial fragmentation that underlie the increasing socialization of labor under conditions of advanced automation. As Anna Tsing remarks, ethnoracial differentiation has come to function as a structuring element in the reproduction of global capitalism, not merely as a "decoration" of it.[116] As the case of the Atacama Desert reveals, the systematic commodification of labor-powers of heterogeneous complexity has thus enabled the functional integration of the supply chain of extraction, rendering its own spatial register through an expanding yet fragmented and geographically uneven fabric of urbanization.

Specifically, shifting trends in industrial organization across the mining industry show that a deeply layered, morphologically variegated built environment has been brought into existence to support and facilitate the exploitation of each organ of the collective laborer in accordance with the productive and ethnoracial attributes it embodies. Technopolises and global cities attract the engineers who operate the complex technological systems of mines and mineral-processing facilities, the scientists who codify the knowledge incorporated into such systems, and the financiers who garner the liquidity required by mining firms. Overcrowded campsites, temporary settlements, and polluted towns host subcontracted, low-skill *faeneros* and other industrial laborers of the mining supply chain; *campamentos*, shantytowns, and other precarious settlements house the racialized surplus populations that provide cheap services to workers and companies in mining towns and act as the "labor reserve army" of the mining industry.

---

116   Tsing, "Supply Chains and the Human Condition."

In developing a spatialized reading of an evolving, internally dynamic mining proletariat, this chapter has built upon Ekers and Loftus's invitation to revitalize the production-of-nature thesis by superseding Smith's idea of "the laborer" in the abstract.[117] This, according to such authors, requires us to ceaselessly historicize the processes and relations that bear on the production of nature and to reveal the making of "gendered, raced and classed subjects operating within historically and geographically specific divisions of labor."[118] After all, at the heart of the internal polarization underlying the territorial distribution of the collective laborer of large-scale mining is the tendency Postone observed: increasing the productive power of the working whole comes at the expense of the productive power of the individual. This "despotism of the collectivity," which, for Postone, is structured by considerations of productivity and efficiency, is effected at the cost of the individual worker.[119] With the development of the working organism that is the collective laborer, Marx notes how the individual is fragmented and transformed into a "crippled monstrosity" by being forced into a particular skill at the expense of a "world of productive drives and inclinations."[120]

The mode of human universality assembled by capital elucidates as socially grounded a rich specificity in terms of a vast multitude of concrete labors and social relations. This, according to Postone, points toward the possibility of *another universalism*, one based not on abstraction but on concrete specificity. The Herculean forces of production conjured by capital accumulation on a world scale, Postone asserts, can also give rise to the historical possibility that people can claim what is produced socially in alienated form.[121] In the case of primary-commodity production, the real subsumption of peasant communities to capital—depeasantization—has created its own countertendency toward (re)peasantization. This new iteration of peasant experience and practice, however, is premised not on a reactionary attachment to the idyllic farm of the past, but on vibrant synergies and dialogues with modern

---

117   Michael Ekers and Alex Loftus, "Revitalizing the Production of Nature Thesis: A Gramscian Turn?" *Progress in Human Geography* 37 (2012): 1–19.

118   Ibid., 7.

119   Postone, *Time, Labor, and Social Domination*, 333.

120   Marx, *Capital*, Vol. 1, 481.

121   Postone, *Time, Labor, and Social Domination*, 162.

science, militant environmentalism, and labor movements.[122] As McMichael has noted, this budding form of *campesino* politics "represent[s] the possibility of a peasant modernism, dedicated to an 'agrarian citizenship,' via politics of ecology and food sovereignty anchored in an episteme of politically reconstituted place."[123]

For these reasons, the very possibility to imagine and collectively build an *alter-globalism* that is sensitive to human and ecological needs entails doing away with teleological interpretations of history, where progress and growth are considered the only pathways to emancipation. A truly radical and emancipatory political project will not be universal but pluriversal, a unity of many worlds. The sheer sociocultural heterogeneity of the global working class expresses the urgency for such a shift of vision. As the last two chapters of this book explore in detail, Marx began to grapple with this question late in his life, becoming seriously invested in studying non-Western and ancient forms of communal life so that he could experiment with the idea of the multilinearity of history. Unfortunately, the present state of things negates the possibility for a pluriversal moment of worker solidarity. Amid unprecedented material interdependence in the organization of capitalist production, the global working class has been traversed by diverse forms of material and ideological fragmentation. The evolving forms of subaltern fragmentation and struggle will be dealt with in chapter 7. Chapter 4 reveals how an emphasis on movement, speed, and connectivity has transformed the spaces of primary-commodity production in recent years.

122   Raj Patel, "International Agrarian Restructuring and the Practical Ethics of Peasant Movement Solidarity," *Journal of Asian and African Studies* 41, no. 1/2 (2006): 71–93; Raj Patel, *Stuffed and Starved: The Hidden Battle for the World Food System* (New York: Melville House Publishing, 2012 [2007]), chapter 7; McMichael, "Peasant Prospects in the Neoliberal Age"; Miguel Altieri and Víctor Manuel Toledo, "The Agroecological Revolution in Latin America: Rescuing Nature, Ensuring Food Sovereignty and Empowering Peasants," *Journal of Peasant Studies* 38, no. 3 (2011): 587–612.

123   McMichael, "Peasant Prospects in the Neoliberal Age," 408.

4. CIRCULATION

State Power and the Logistics Turn in the Extractive Industries

The hole is an essential feature of the extractive landscape, but the hole is just the start.[1]

Covenants, without the sword, are but words?[2]

Introduction

Technical sabotage, paradigmatically manifested in the blockade, has gradually emerged as one of the chief tactics of sociopolitical mobilization in the twenty-first century. Initially popularized during a cycle of labor strikes in France in 2006, the slogan "Power is logistic. Block everything!" has acquired wide-ranging resonance in recent years as the chokepoints of global supply chains increasingly assert themselves as key sites of political contestation.[3] In Latin

---

1   Bridge, "Hole World."
2   Thomas Hobbes, cited in Hannah Arendt, *On Violence* (New York: Harcourt, 1970).
3   Brett Neilson, "Five Theses on Understanding Logistics as Power," *Distinktion* 13, no. 3 (2012): 323–40; Invisible Committee, *To Our Friends*, chapter 3; Deborah Cowen, "A Geography of Logistics: Market Authority and the Security of Supply Chains," *Annals of the Association of American Geographers* 100, no. 3 (2010): 600–620; Cowen, *Deadly Life of Logistics*; Charmaine Chua, "The Container: Stacking, Packing, and Moving the World," *Funambulist* 6 (2016): 41–45.

America, attacks on infrastructure have arguably become the most common tactic in the struggle against the operations of extractive industries. Such tactics include sabotaging oil pipelines, blocking roads and railways, occupying oil rigs, and burning firms' cargo trucks, machinery, and other fixed assets. In Chile, the strategic relevance of transnational logistical systems for primary-commodity production was first cast into stark light in 2012, after local communities blocked for five days the highways and railways that connect the port of Huasco—one of the largest in the Atacama Desert—to its surrounding resource hinterlands. Huasco and neighboring villages, having widely broadcast setting barricades on fire and clashing with antiriot police amid tear gas, attracted media attention not because they were shutting down the operations of a few local mines but because they were jeopardizing an entire circulatory system of raw materials that stretches well into Asia and other parts of the world.[4] At stake in these blockades were the activities of power plants and mines but also of smelting facilities, port infrastructures, factories, construction sites, and whole fleets of cargo trucks, trains, bulk carriers, and container ships.

These emerging expressions of political struggle indicate that the metabolic system of primary-commodity production becomes concretely actualized in the shafts and pits where minerals are first wrested from the subsoil, as well as in multifarious logistical flows spanning oceans and continents. However, as noted in the introduction, most scholarly accounts of resource extraction tend to construe their object of study in a static and self-contained manner. This often leads them to grant analytical prevalence to the site of extraction in its own right, not to how and why the site comes into being in the first place. Throughout modern history, spaces of primary-commodity production have been entangled in dense webs of connectivity that meld them together with distant geographies of political rule, finance, urbanization, manufacturing, and information. In fact, the material separation between spaces of extraction and spaces and manufacturing, as Bunker and Ciccantell show, is what has triggered the technological breakthroughs in transportation that have come to define

---

4   Arboleda, "In the Nature of the Non-City."

historical cycles of accumulation.[5] However, as Bridge rightly suggests, the debate on resource peripheries has advanced through a default to *national-scale* modes of analysis that pushes questions about the transnational organization of production into the background.[6]

Through exploring port cities in northern Chile, this chapter assesses the territorial and sociopolitical implications of the logistics turn in the extractive industries. The systematic adoption of organizational imperatives toward flow, connectivity, and speed in mining operations has led a modality of logistical urbanization in which the governance of mineral flows assumes increasing salience vis-à-vis the moment of extraction. This evolving mode of connectivity urbanism has underpinned the rise of a transpacific logistical corridor that weaves together mineral deposits in the Chilean Andes with port and manufacturing cities in East Asia. To the extent that maritime trade in raw materials intensified substantially after the turn of the twenty-first century as a result of increasing demand from Asian economies, the mining industry faced substantial pressures to reduce transport costs and increase the circulation speed of minerals. New technological innovations have been harnessed to attain functional integration between the phases of the mining process (i.e., forecasting, blasting, shoveling, haulage, crushing, and processing) and between the mining, port, and shipping industries as a whole.

As the Latin American mining industry shifts its emphasis from operations on site to the "global value chain," it has made many organizational efforts at various spatial scales to reduce turnover times of capital and hence attain deeper integration with maritime infrastructural corridors and spaces of advanced manufacturing in Asia.[7]

5   Stephen Bunker and Paul Ciccantell, "Generative Sectors and the New Historical Materialism: Economic Ascent and the Cumulatively Sequential Restructuring of the World Economy," *Studies in Comparative International Development* 37, no. 4 (2003): 3–30; Bunker and Ciccantell, *Globalization*.

6   Gavin Bridge, "Global Production Networks and the Extractive Sector: Governing Resource Based Development," *Journal of Economic Geography* 8 (2008): 389–419.

7   World Bank, *Benchmarking Container Port Technical Efficiency in Latin America and the Caribbean: A Stochastic Frontier Analysis*, Sustainable Development Department, World Bank, 2013; ECLAC, *El gran desafío para los puertos: la hora de pensar una nueva gobernanza portuaria ha llegado*, Department of Natural Resources and Infrastructure, Economic Commission of Latin America and the Caribbean,

Simultaneously to this overhaul in the materiality of the labor process, it has been argued that the post-2008 context has brought with it a more coercive, centralized, and militaristic configuration of state power.[8] Accordingly, this chapter also highlights the fact that the emergence of logistical urbanization has entailed contradictory and yet complementary tendencies toward the internationalization and concentration of the political authority of the late liberal state. To conceptualize the political mediations that animate this complex, transoceanic system of logistical connectivity, I build upon a rich yet overlooked tradition of Marxist state theory that has uncovered the nature of the state as a fetishized form of existence of capitalist relations of production.[9]

In general terms, this chapter deciphers the sociospatial ramifications of the mining industry embracing the imperatives of flow, homeostasis, and connectivity that underpin logistics as an overarching "science of systems."[10] It begins by laying out the ways in which a logistics turn in the extractive industries has spearheaded relevant sociotechnical reconfigurations in the mining, port, and shipping industries in Chile and beyond. The second section reflects on the evolving modes of governance and state authority that have resulted from the revolutions in surplus value production brought about by an emphasis on speed. To do this, I briefly highlight the

2015; Asia-Pacific Economic Cooperation (APEC), *Case Study on the Role of Services Trade in Global Value Chains: Transport Services in Chile*, Asia-Pacific Economic Cooperation, 2016; APEC, "Exploration on Strengthening of Maritime Connectivity," Transportation Working Group, Asia-Pacific Economic Cooperation, 2016; *Financial Times*, "Copper Concentrate Supply Hit by Disputes in Chile and Indonesia," March 12, 2017, ft.com.

8   Bonefeld, "Adam Smith and Ordoliberalism"; Bonefeld, *Critical Theory and the Critique of Political Economy*; Werner Bonefeld, *The Strong State and the Free Economy* (Lanham, MD: Rowman & Littlefield, 2017); Ian Bruff, "The Rise of Authoritarian Neoliberalism," *Rethinking Marxism* 26, no. 1 (2014): 113–29; Jordan Camp and Christina Heatherton (eds.), *Policing the Planet: Why the Policing Crisis Led to Black Lives Matter* (New York: Verso, 2016); Jeremy Roos, "Managing Disorder," *ROAR* 4 (2017), https://roarmag.org/magazine/managing-disorder.

9   John Holloway and Sol Picciotto (eds.), *State and Capital: A Marxist Debate* (London: Edward Arnold, 1978); Braunmühl, "On the Analysis of the Bourgeois Nation State"; Holloway, "State and Everyday Struggle"; Clarke (ed.), *State Debate*; Bonefeld et al. (eds.), *Open Marxism*; Bonefeld, *Critical Theory*; McNally, "Blood of the Commonwealth."

10   Cowen, *Deadly Life of Logistics*, chapter 1.

theoretical contributions of Marxist approaches to the state from a form-analysis perspective, and their potential to inform current trends and transformations in the organization of state power. The final section explores port cities in northern Chile to elucidate some of the central elements of logistical urbanization and its concomitant expressions of state authority, sociopolitical revolt, and ecological degradation.

The Circulation of Capital and the Logistics Turn in the Extractive Industries

The development of abstract time laid the foundation, perhaps even more decisively than the command over physical space, for the historical determination of capital as an impersonal yet objective form of social domination.[11] The Marxian labor theory of value derives its significance from the measurement of human exertion enabled by fragmenting time into homogeneous, quantitative units—a historically unique mode of temporality that was alien to premodern societies. As a direct precondition for the expansion of value, the act of shortening the turnover times of capital has, from the outset, been one of the major drivers for technological innovation under capitalism. The development of faster, cheaper, more efficient means of transportation has been a key determinant of quantum leaps in the organizational dynamics of social forms of labor and spatial configurations to the built environment. Although production and consumption cannot take place simultaneously, shortening the timespan that separates the two has pushed modern science and engineering to strive endlessly to intervene in human and extrahuman lifecycles—from biological reproduction in farm animals to irrigation-system design in agriculture to propulsion systems in bulk carrier vessels.

In volume 2 of *Capital*, which is devoted almost in its entirety to the question of circulation, Marx considers that the more production comes to rest on exchange value, hence on exchange, the more important the physical infrastructures of trade—the means of communication and transport—become for the costs of

---

11   Postone, *Time, Labor, and Social Domination*, chapter 5.

circulation.[12] By its very nature, says Marx in the *Grundrisse*, capital drives beyond every spatial barrier and thus "the creation of the physical conditions of exchange—of the means of communication and transport—the annihilation of space by time—becomes an extraordinary necessity for it."[13] As Bunker and Ciccantell's study of natural-resource frontiers in world-historical perspective demonstrates, far from being accessory to it, primary-commodity production has been the main breeding ground for innovations in maritime and ground transportation.[14] As the material separation between resource peripheries and manufacturing and financial centers increased, modern economic powers faced consistent pressure to reduce transport costs in order to achieve trade dominance. Building larger ports, ships, bridges, railways, and other infrastructural elements, has therefore been internal to the business of extraction throughout the *longue durée* of capitalism. Chapter 2 analyzed the major technological innovations introduced by economic powers to secure access to raw materials throughout the phases of capitalist development, so the focus here is to excavate the peculiar nature of circulation concomitant with extraction during contemporary conditions of production.

Grasping the role that the logistics revolution has performed in the transnational organization of resource extraction demands that we understand circulation in a double sense. That is, circulation involves the actual, physical movement of commodities, labor, and instruments of production across diverse infrastructural technologies, such as pipelines, transmission lines, highways, airports, etc. But most importantly, circulation also needs to be understood as the successive movement of capital through its different phases, or *modes of existence*. In the M-C-M' circuit, Marx contends, "value suddenly presents itself as a self-moving substance which passes through a process of its own, and for which commodities and money are both mere forms."[15] The realization of capital hence involves transitioning from one of its phases or forms of appearance (money, forces of production, labor) to

---

12   Marx, *Capital*, Vol. 2.

13   Karl Marx, *Grundrisse: Foundations of the Critique of Political Economy* (New York: Penguin, 1973).

14   Bunker and Ciccantell, *Globalization*.

15   Marx, *Capital*, Vol. 1, 256.

another. For money to become determined as forces of production, bankers and financial intermediaries need access to adequate telecommunication and information systems; for forces of production to become determined as commodities, discipline has to be enforced on the shop floor and fixed capital has to function smoothly and swiftly; for commodities to become determined as profit, efficient distribution and marketing mechanisms need to be put in place. On that basis, the ability to increase the speed at which capital transitions through its phases is what gives firms an upper hand in a setting of intercapitalist competition.

Although circulation has been a matter of necessity for the reproduction of capital since its genesis, the qualitative difference in the present cycle of accumulation is that the logistics revolution—in the interest of speed—has deliberately and decisively blurred the boundaries between making and moving—production and distribution.[16] At stake here is a tendency to accelerate the valorization of capital and to integrate the phases of production conceptually and organizationally under systemic and holistic modes of corporate rationality. Underpinned by new methods for rational calculation, simulation modeling, and nonlinear programming that emerged at the onset of the Cold War, companies developed a "total-cost" approach that allowed them to aggregate a wide array of functions previously understood as alien to mere production. "Inventory carrying and obsolescence, warehousing, transportation, production alternatives, communications and data processing, customer service, alternative facilities use, channels of distribution, and cost concessions," in Cowen's words, began to be calculated as part of an integrated profit-maximization scheme under the total-cost rationale.[17] The rise of integrated distribution management implied a fundamental conceptual shift in the role of transport within the overall architecture of production. Once considered a cost, transportation began to be understood as a privileged site for adding value to commodities.[18]

The world-historic significance of the logistics revolution in the

---

16   Sheller, "Aluminum Dreams"; Cowen, *Deadly Life of Logistics*; Sandro Mezzadra and Brett Neilson, "Operations of Capital," *South Atlantic Quarterly* 114, no. 1 (2015): 1–9.

17   Cowen, *Deadly Life of Logistics*, 36.

18   Ibid.

generalized dynamics of profit-making is of such magnitude that some argue that instead of industrial powerhouses or world factories, Asian Tigers (especially Japan, China, and South Korea) are more adequately conceptualized as "logistics empires."[19] With more than 2,000 vessels, China's massive merchant-marine fleet is without parallel; its fishing armada boasts some 200,000 trawlers. Five of the top ten container ports in the world are located along China's mainland coast. In 2015, the five big Chinese carriers together controlled 18 percent of all container shipping handled by the world's top twenty companies— higher than Denmark, the second in line.[20] In this light, the reorganization of the mining industry into global supply chains cannot be understood in isolation from the massive tectonic shift that has rendered the Pacific Ocean the main infrastructural corridor for world trade. This newly configured geoinfrastructural apparatus is the result of sprawling maritime trade routes connecting East Asia with the rest of the world. It is also the product of a process of modernizing and upgrading airport infrastructure, air freight, and commercial flight capacity across the cities of the Pacific Rim. It has been argued that the geography of aviation, generally considered, is shifting eastward as airports in the Asian continent attract increasing numbers of passenger and cargo traffic.[21]

The figures in maritime cargo traffic illustrate this geoeconomic transformation. In the 1970s, transatlantic shipping represented 80 percent of global trade, whereas by 2013 it accounted for only 40 percent, with the majority of maritime commercial flows already taking place across the Asia–Middle East–Africa–Americas axis.[22] The genesis of this reconfiguration in naval capacity, Bunker and Ciccantell explain, can be traced to Japan and South Korea's need to reduce transport costs to

19   Ciccantell, "China's Economic Ascent"; KPMG, *On the Move in China: The Role of Transport and Logistics in a Changing Economy*, 2011; Cowen, *Deadly Life of Logistics*; Khanna, *Connectography*.

20   Khanna, *Connectography*; *Financial Times*, "How China Rules the Waves," January 12, 2017, ft.com.

21   John Kasarda, "The Evolution of Airport Cities and the Aerotropolis," in John Kasarda (ed.), *Airport Cities: The Evolution* (London: Insight Media, 2006); John Bowen, "Continents Shifting, Clouds Gathering: The Trajectory of Global Aviation Expansion," in Lucy Budd, Steven Griggs and David Howarth (eds.), *Sustainable Aviation Futures* (Bingley, UK: Emerald Group, 2013).

22   Khanna, *Connectography*, 242; OECD, *Logistics Observatory for Chile*, 9.

access remote resource peripheries.[23] Quantum leaps in steelmaking and ship propulsion techniques allowed Japan to increase tanker capacities from 20,000 dwt in 1950 to 300,000 dwt in 1968, a development that allowed it to become the major importer of raw materials by 1984.[24] In 1995, China became the world's largest steel producer, which set it on the path to become the main importer of raw materials and develop its colossal maritime fleet.[25]

The steelmaking and shipbuilding prowess initially achieved by Japan and South Korea materialized a tenfold increase in the number of dry-bulk carriers (cargo ships built to carry minerals and grains) between 1961 and 1992 from 471 to 4,846.[26] In Latin America, container traffic more than doubled after the turn of the millennium, as Asia gradually became a major destination for its raw materials exports. The volume of containerized cargo in the region went from 17 million twenty-foot equivalent units (TEUs) in 2000 to 40 million in 2010, with an average compound annual growth rate of 10 percent.[27] Even more epochal, perhaps, are the mining industry's reconfigurations to deepen its integration with the rest of the supply chain. According to Chile's Ministry of Transport and Telecommunications,[28] dry-bulk cargo (mainly mineral concentrate) amounts to 60 percent of all national exports and 37 percent corresponds to biomass (mainly cellulose, foodstuffs, fruits). The incentives for increasing the efficiency of these flows are therefore substantial and have exerted a strong organizational push toward viewing primary-commodity production and transport as an integrated whole. Antofagasta Plc., one of the major mining companies operating in northern Chile, indicates that 14 percent of its weighed average cash costs in 2011 were in shipping and tolling charges.[29]

Codelco, which is Chile's main state-owned mining company and one of the major copper producers in the world, has been

23   Bunker and Ciccantell, *Globalization*.
24   Ibid., 193.
25   Ciccantell, "China's Economic Ascent."
26   Ibid., 196.
27   World Bank, *Benchmarking Container Port Technical Efficiency*, 2.
28   Ministry of Transport and Telecommunications, *Plan Nacional de Desarrollo Portuario* (Santiago: Government of Chile, 2013), 16.
29   OECD, *The Competitiveness of Global Port Cities: The Case of Antofagasta-Chile*, OECD Regional Development Working Papers, 2013, 15.

developing efforts to integrate the phases of mining since 2007 through its "Codelco Digital" initiative.[30] The company began to realize that a compartmentalized approach to corporate processes was undermining production turnover times. Technological advances in remote sensing, geospatial modeling, and control systems have allowed it to introduce sensors, AI, and statistical learning methods to the whole spectrum of operations (forecasting, blasting, haulage, leaching, lab analysis, smelting, refining), thus reducing the communication gap between them. Most importantly, operational redesign processes have not been confined to mineral extraction and processing. Mineral buyers in Asian markets have been increasingly requesting that the mining industry implement technical systems to make products traceable throughout the entire supply chain—that is, the "history" of the mineral, from the point of its extraction to its delivery in the port of destination. This allows the ore-smelting, alloy, manufacturing, and construction industries to identify more accurately the particular attributes of the mineral, differences among batches, and impurities, among other relevant details.

In implementing technologies for mineral traceability and supply-chain mapping, the mining, shipping, ground-transportation, and port industries have had to standardize operations and technological systems as well as share information on an ongoing basis.[31] Chinese, Japanese, and South Korean ports, for example, have begun to lead systematic efforts to computerize port operations and implement cloud computing, remote sensing, and big data to integrate payment, warehousing, logistics, and customs clearance.[32] Although mineral traceability began as a private initiative to reduce costs, it has recently been embraced by civil society and multilateral organizations as a way to increase transparency and avoid human-rights violations, especially with respect to

---

30   Codelco, "Codelco Digital," March 3, 2011, codelco.com.
31   Interview with a Chilean mining executive, January 10, 2017.
32   APEC, *Final Report: Automated Transport Management Systems Implementation for Optimizing Logistics within the Asia-Pacific with an Emphasis on ITS and GNSS Application*, APEC Transportation Working Group, Asia-Pacific Economic Cooperation, 2012, 20; APEC, *Case Study on the Role of Services Trade*, 22.

conflict minerals such as coltan, tin, tungsten, and gold.[33] In Chile's extractive industries, the new managerial emphasis on operational integration has been so systematic that a recent report from the Asia-Pacific Economic Cooperation (APEC) forum, states that the involvement of these industries in the global value chain (GVC) "makes clear that transport can also be a source of real value addition in the global value chain context."[34] Traditionally, the report claims, transport services have been seen primarily as a cost component of production. Chile's liberalization of key transport modes during recent years has overturned this assumption, because transport contributed to increase GVC participation—as measured by domestic value added—by around 7 percent.[35]

To cope with the expanding scale of primary-commodity production, the Latin American port industry has likewise set in motion an aggressive process of technological and organizational modernization. Increasing cargo-handling operational capacity and transfer speed has been the main target of this endeavor. In Chile, automation and mechanization (including the acquisition of Gottwald, Liebherr, and Gantry cranes between 1995 and 2002) have enabled ports to increase transfer speed by a factor of three.[36] As in the mining industry, there is a clear tendency toward vertical reintegration in large-scale cargo operators, which encompass ports, ground transport, and maritime shipping companies. Buffer zones for logistical backup to port facilities that require reducing congestion have also been developed, mostly to increase cargo-handling speed.[37] The assumption that these revolutions in surplus-value production have been contingent upon minimal state intervention and aggressive deregulation has become received wisdom among free-trade advocates and even relevant strands of critical scholarship. As we will see in the next section, however, appearances are often deceitful.

---

33   OECD, *OECD Due Diligence Guidance for Responsible Supply Chains of Minerals from Conflict-Affected and High-Risk Areas*, 2nd ed., 2013.

34   APEC, *Case Study on the Role of Services Trade*, 6.

35   Ibid., 36.

36   Fundación Chile, *Trabajadores portuarios de Chile: Entre la precariedad y la esperanza* (Santiago: Fundación Chile, 2016), 17.

37   Ministry of Public Works, *Infraestructura portuaria y costera Chile 2020* (Santiago: Government of Chile, 2009).

Transoceanic Corridors and the Form of the State

The emergence of logistics as one of the most relevant questions of the twenty-first century has also begotten a spatialized version of the type of hyperglobalist ideologies that marked the so-called "globalization debate" in the 1990s. The astonishing scale of the transcontinental logistical infrastructures for the acceleration of world trade, it is usually argued, dramatically clashes with inherited geopolitical configurations of the Westphalian model of territorial and national sovereignty. Most importantly, these sociotechnical arrangements are considered to be not only of a higher order than state power but also ontologically alien to it. In Parag Khanna's words, "supply chains and connectivity, not sovereignty and borders, are the organizing principles of humanity in the twenty-first century."[38] In the epochal, world-historic movement of human civilization from geography to what Khanna terms *connectography*, organizing the world according to political space is giving way to organizing it according to functional space. Just as the world evolved from vertically integrated empires to horizontally interdependent states, he contends, it is now "graduating toward a global network civilization whose map of connective corridors will supersede traditional maps of national borders."[39]

In a somewhat more nuanced yet similar methodological register, Keller Easterling's notion of *extrastatecraft* is meant to designate the transnational infrastructural technologies whose governance is, to a large extent, beyond the rule of law.[40] Thus, in being "far removed" from familiar legislative processes, extrastatecraft generates "de facto forms of polity faster than even quasi-official forms of governance can legislate them."[41] What is most remarkable about this emerging rubric of "spatial hyperglobalism" is perhaps that, much like its preceding theoretical iterations, it spans the whole political spectrum. Despite the divergence in politico-normative orientations, radical social theory has tended to construe the political implications of the logistics revolution in a very similar light. Brett Neilson, for example, suggests that logistical

---

38   Khanna, *Connectography*, 20.

39   Ibid., 6.

40   Keller Easterling, *Extrastatecraft: The Power of Infrastructure Space* (New York: Verso, 2016).

41   Ibid., 15.

corridors interrupt the continuity of state territories and normativity and that, in so doing, create the conditions for an era in which "economic and state sovereignty has been disaggregated to an extent even neoliberal theorists did not anticipate."[42] The Invisible Committee is even more emphatic about the radical disconnection between economics and politics that allegedly lies at the heart of the logistical systems of the current industrial era. Power, they argue, has recently come to have an architectural and impersonal, not representative or personal, nature.[43] Rather than in polities or in sovereign governmental organizations, they contend that the locus of political authority now resides in the materiality of logistical systems.

There is, however, an element of truth in the Invisible Committee's idea of power as being more logistical than institutional. Without a doubt, the expansion of transcontinental corridors of intermodal transport that weave together railways, highways, automated port infrastructures, and supertankers has further intensified the impersonal, directionally purposed compulsions that are immanent to the reproduction of capital as an alienated subject. Nevertheless, these impersonal compulsions do not exist on a separate plane of reality from the nation-state; neither are they juxtaposed with the political mediations of sovereign rule. Grasping the essential unity of what appear as distinct or disaggregated spheres of social reality demands a theory of the state that starts from the actual conditions in which the human life process asserts itself. Such was the aim of an overlooked intellectual tradition of Marxist scholars that emerged around the 1970s and whose work unfolds a critique and reconceptualization of the capitalist state, especially through the framework of *modes* or *forms of existence*—rather than through "structures," essences, or first principles.[44]

---

42 Neilson, "Five Theses," 335.

43 Invisible Committee, *To Our Friends*, 83.

44 For an overview, see Holloway and Picciotto, *State and Capital*; Clarke (ed.), *State Debate*; Bonefeld et al. (eds.), *Open Marxism*. The fragmentation of the capital relation into distinct political and economic spheres, Holloway argues, is perhaps the most important aspect of commodity fetishism. It obfuscates the total experience of class domination by fragmenting the individual into citizen on the one hand, and laborer on the other. See Holloway, "State and Everyday Struggle." A materialist theory of the state therefore sets out from the fundamental unity of human experience in order to grasp the categories of the modern state as illusory, yet objective forms of everyday bourgeois class practice. The contradictory and yet unitary nature of these categories, to

Taking seriously the young Marx's claim that science should have sensuous practice as its starting point, authors in the above-mentioned intellectual tradition were dissatisfied with the contending visions of what was branded the "Miliband-Poulantzas" state debate in 1970s Britain.[45] On the one hand, they were critical of the "vulgar materialism" of Ralph Miliband, where the state is considered a mere instrument or "executive committee" of the capitalist class—an approach whose lineage can be clearly traced to the Second International and to the Soviet "Diamat." On the other hand, authors in the "form-analysis" tradition were also interested in challenging the structural-functionalist approach to the state that informed the work of Nicos Poulantzas, which stressed the "relative autonomy" of the state from the domain of capital accumulation—and thus of class struggle.[46] By contrast to the structural determinism and political reductionism prevalent in both sides of the debate, a "form-analysis" view of the state rejected the separation of politics and economics as a methodological principle. Instead, it proposed to understand the state as an illusory expression or mode of existence of the social relations that underpin it: that is, the material relations of social and ecological life.

In other words, the apparently autonomous nature of the state is, like all forms of fetishism, simultaneously illusory and real.[47] To

---

echo Marx, is clearer to the "popular mind" than to the vulgar economists. See Marx, *Capital*, Vol. 3 (New York: Penguin, 1981 [1894], 956).

45   Marx, *Economic and Philosophic Manuscripts*, 111.

46   Nicos Poulantzas, "The Problem of the Capitalist State," *New Left Review* 58, reprinted in Robin Blackburn (ed.), *Ideology in Social Science* (London: Fontana, 1969); Nicos Poulantzas, *Political Power and Social Classes* (London: New Left Books and Sheed and Ward, 1973). The concept of *relative autonomy*, which Althusser coined in order to explain the status of the political superstructure relative to the economic base, has been considered a contradiction in terms. As Iñigo Carrera explains, aside from being contradictory, the term only expresses a quantitative relation of deviation of the superstructure with respect to the base. The qualitative relation between the two, and therefore the necessity of the superstructure and its determination, remains unsolved in Althusser's elaboration. Juan Iñigo Carrera, "Acerca del carácter de la relación base económica-superestructura política y jurídica: la oposición entre representación lógica y reproducción dialéctica," in Gastón Caligaris and Alejandro Fitzsimmons (eds.), *Relaciones económicas y políticas: Aportes para el estudio de su unidad con base en la obra de Karl Marx* (Buenos Aires: Universidad de Buenos Aires, 2012).

47   Although the forces of production (i.e., machines, computers, trucks) constitute a mere form of capital as it transitions through its phases, their objectivity is concrete.

decipher the real social relations behind the reified forms of appearance of the state, then, one must first interrogate the evolving modalities in which the process of socioecological metabolism is mediated, produced, and organized. An analysis that sets out from the various configurations (forms) assumed by social practice will be adequately positioned to capture the nature of the state as an institutional apparatus in constant co-evolution and codetermination with processes of market-making, technological change, and social production rather than as an ahistorical or suprahistorical black box. Moreover, insisting on the nature of the state as a mode of existence of social practice is fundamental for capturing the essentially global content expressed through the national forms of domestic spheres of accumulation. If, as Marx argues, "the tendency to create the world market is inherent to the concept of capital itself," then the very concept of a national state is better understood as the concentrated expression of a process whose scale is planetary.[48] In Claudia von Braunmühl's words:

> The nation-state is thus not merely the historical form of organization within which capital first develops and grows into a nationally centered complex of production and exchange, it is also . . . an indispensable instrument necessary to secure the profitable outcome of the valorization of national capital in its competition with the many other capitals combined together in nation-states . . . Even if the internationalization of accumulation involves the increasingly international determination of exploitation, and the direction of the particular national production processes are structured by conditions of international competition and differences in productivity, the authority which safeguards this exploitation still continues to be mediated nationally.[49]

The sensuous existence of forces of production, like that of the state, makes them assume an alienated objectivity that can be sometimes hostile to the very process of capital accumulation (Simon Clarke, "State, Class Struggle, and the Reproduction of Capital," in Clarke (ed.), *State Debate*. Just as a machine can break down or get devalued and just as living labor can unionize or shirk, state activity can also depart from the immediate interests of the capitalist class, or fractions of it, at certain points.

48   Marx, *Grundrisse*, 408.
49   Braunmühl, "On the Analysis of the Bourgeois Nation State," 176.

The dialectical reading of the state that this chapter proposes therefore departs from the methodological nationalism that abounds in most studies of resource extraction, as well as from the hyperglobalism of some of the literature on logistics and infrastructure space. Both ends of this spectrum oscillate between a problematic "politicism" (in the former) and "economicism" (in the latter) that obfuscate the structural unity of economic and political mediations in the making of the logistical networks of extraction. As chapter 2 argued, it is not that the world market is a collection or "patchwork" of many national economies. Rather, the world market is organized in the form of national economies as its aliquot parts. Market freedom therefore not only presupposes the political state but it is in fact *premised* on the state as its political authority.[50] If the state is the political form of market liberty, as Bonefeld points out, then the revolution in logistics—a revolution in the production and realization of value—should have developed alongside an equally significant reconfiguration in the institutional materiality or *form* of the state.[51] In fact, very little is known about the institutional mediations and "state rescaling" processes—to use Brenner's formulation"[52] that have been unfolding to facilitate and enhance the functional connectivity of the mining industry across the Pacific Ocean.

The APEC, of which Chile is an active member, is perhaps the most tangible expression of the internationalization of state power and the building of institutional capacity required to conjure a whole new scale of mineral flows. Composed by twenty-one economies of the Pacific Rim, the APEC was launched as an institutional setting for advancing trade liberalization, economic integration, and technical cooperation among its members. The APEC is not a supranational organization; it functions as a forum for member economies to discuss themes that are transversal to their domestic agendas. Its declarations, for that reason, are not legally binding. The silent compulsion of market forces, however, compensates for the lack of legal obligation, because member economies are often unrelenting in implementing the forum's recommendations. A recent "Connectivity Blueprint" report by the APEC highlights

---

50   Bonefeld, "Adam Smith and Ordoliberalism."
51   Ibid.
52   Brenner, *New State Spaces.*

that GVCs have become a dominant fact of the global economy,[53] and member economies have eagerly embraced them as a strategic priority in their policy landscapes "at all levels of development." In addition to reiterating its efforts to strengthen supply-chain efficiency, the document resolves to strengthen "physical connectivity," "institutional connectivity," and "people-to-people connectivity."[54] Because the APEC economies have identified differences in their labor regimes as a key factor hampering the productivity of ports and automation processes,[55] an APEC Transportation Working Group (TWG) was established to advance the standardization of policy and legal systems.

One of the APEC economies' key objectives in the Bogor Declaration of 1994 was a 5 percent reduction in transaction costs within ten years. This, it was argued, required the amendment of legal and administrative procedures that "impede and delay movement of goods and increase general cost of goods movement within the region."[56] The APEC thus established a Supply Chain Development Initiative in Singapore in 2009, which has become one of its main transport and logistics initiatives. Moreover, at the 2012 APEC summit, foreign affairs and trade ministers issued a joint declaration concerned with the specific political initiatives to enhance supply-chain performance, which is surprisingly evocative of "total cost analysis," declaring that

> supply chains should be considered as a single modern network equipped with smart technologies, including intelligent transportation systems (ITS), monitoring systems based on Global Navigation Satellite Systems (GNSS), automated cargo identification systems based on RFID (radio frequency identification), and automated transport management logistical services.[57]

The Alianza del Pacífico (Pacific Alliance) is a multilateral initiative comprised of the governments of Chile, Colombia, Mexico, and Peru to facilitate material integration across the Pacific Ocean. Launched in 2011, its main objective is to build an area of "deep integration"

---

53   APEC, *Case Study on the Role of Services Trade.*
54   Ibid.
55   Ibid.
56   Ibid., 5.
57   Ibid., 6.

that can move progressively toward the free movement of goods, services, resources, and people. Its working groups include Trade and Integration as well as Mining Development, both of which work to reduce technical and institutional barriers to trade and the flow of primary commodities more generally. Among the core objectives of the Mining Development working group are "the integration of value chains" and "the development of scientific and technical capabilities to drive growth."[58] Just by scratching the surface of these evolving modes of statecraft, it is possible to capture in full the implications of Bonefeld's claim that economic liberty (and the territorial infrastructures that support it) has no independent reality but is "a practice of government."[59] The idea of minimal state intervention as the underlying principle of neoliberal governance, however, is deeply entrenched in theory—even among practitioners in state agencies and mining companies. In fact, state activity at all spatial scales has been aggressively geared toward strengthening the material and institutional configurations of the integrated logistical systems for extraction.

In addition to regionally specific initiatives for economic integration, multilateral organizations such as the ECLAC and the Organisation for Economic Co-operation and Development (OECD), have also engaged in concrete efforts to monitor, implement, and standardize governance practices and protocols that can strengthen performance in logistics. From its Maritime and Logistics Profile, the ECLAC has been arguing for a shift from a unimodal to a "systemic and integrated" approach to port planning and design that includes integration with the hinterland, logistical systems, production, and transport.[60] In a 2016 report, the OECD's International Transport Forum concluded that Chile lacked key performance indicators to measure transport and logistics operations and recommended creating a logistics observatory to bridge this gap in data availability.[61]

Zooming in to the national scale, a similar pattern of dynamic

58   Pacific Alliance, alianzapacifico.net.
59   Bonefeld, *Critical Theory*, 176–77.
60   ECLAC, *El gran desafío*, 3.
61   OECD, *Logistics Observatory for Chile*.

institutional redesign and state restructuring becomes clearly discernible in Chile. The focus of these efforts encompasses several state agencies and centers on the principles of logistics—such as connectivity, intermodality, integration, automation, supply chains, and efficiency. In 2010, Chile's Ministry of Transport and Telecommunications (MTT) launched the Logistical Development Program, an institutional platform commissioned to lead planning in intermodal transport systems and physical infrastructures of connectivity. Since maritime cargo in Chile accounts for 95 percent of all foreign trade (with dry-bulk cargo making up around 60 percent of all exports) and the port system is of strategic relevance for the country's energy and mining industries, the MTT issued a "National Plan for Port Development" in 2013, intended to transcend an erstwhile fragmentary approach to territorial design and adopt what the ministry terms "a sectoral orientation."[62] Adopting such an orientation is urgent because port activity has been evolving toward a wider array of operations. The port's role, according to the MTT, now also concerns integrating processes from the cargo's point of origin to its destination. This particular scenario, the aforementioned plan states, "demands broadening [the port's] functionality toward the territory and conceptualizing it as a relevant link in the coordination of the logistical systems that support its role with respect to the hinterland (or area of influence) they assist."[63] This is the latest, most advanced iteration in an already frantic succession of major reforms to the port industry.

Law 19.542, explicitly intended to "modernize" the national port system, can be considered phase zero in the institutional overhaul that gradually led to the current interagency emphasis on the global supply chain. This law was sanctioned in 1997, when Chile's exports were being systematically reoriented toward Asia and the expansion of trade corridors demanded an upgrade in port efficiency. It privatized the vast majority of port facilities (following the "landlord system" of port regulation) and created an elaborate array of incentives for a highly competitive structure of logistical operations. This period saw the beginning of an aggressive nationwide movement toward automation,

---

62 Ministry of Transport and Telecommunications, *Plan Nacional*.
63 Ibid., 12.

containerization, and standardization of port activity. Ten years later, port fees had been reduced by 30 percent and investment in port infrastructure and technology had surpassed $341 million. Transfer speed had increased by 51 percent and port efficiency by 100 percent.[64] These dazzling economic figures have exerted substantial transformations upon the built and unbuilt environments of major mineral-exporting zones in Chile.

Logistical Urbanization in Northern Chile

Some commentators have recently argued that the post-2008 context has marked a turning point in the evolution of state power and political authority, as the late neoliberal state has leaned toward an ostensibly more authoritarian, coercive configuration. Faced with a loss of legitimacy following financial meltdowns, structural unemployment, and economic recessions, variegated efforts have been deployed to insulate institutional design and policymaking from social and political dissent.[65] This emerging form of "authoritarian neoliberalism," Bruff argues, is less interested in neutralizing resistance and contestation via concessions and forms of compromise, favoring instead the explicit exclusion of subordinate social groups through the constitutional, legally engineered disempowerment of nominally democratic institutions. The burgeoning shift toward increasing the institutional concentration of state authority has—paradoxically—been one of the key tenets or preconditions for the functional integration that has been unfolding across the global economy.

The tendencies toward advanced horizontal integration of industrial systems and toward increasing centralization of state power have exerted their own territorial imprint in the form of what has been termed *logistical urbanism* or *connectivity urbanism*.[66] This emerging formation of

64   Ministry of Public Works, *Infraestructura portuaria*, 11.

65   Neil Brenner, Jamie Peck, and Nik Theodore, "Neoliberalism Resurgent? Market Rule after the Great Recession," *South Atlantic Quarterly* 111, no. 2 (2012): 265–88; Bruff, "Rise of Authoritarian Neoliberalism"; Bonefeld, *Strong State*.

66   Markus Hesse, *The City as a Terminal: The Urban Context of Logistics and Freight Transport* (Hampshire, UK: Ashgate, 2008); Cowen, *Deadly Life of Logistics*; Caroline Filice Smith, *Logistics Urbanism: The Socio-Spatial Project of China's One Belt,*

territorial planning and design has been contingent on the production of an entire technological landscape that is functional to the organization of industrial systems for transnational connectivity and the circulation of goods. Because urban space is intrinsically congested, politically contested, and riddled with unforeseen events, logistical urbanism combines a wide array of architectural, technical, institutional, and disciplinary means to mitigate and avoid any obstacles that might undermine the smooth functioning of logistical circulation. The containerization of cargo, for example, evolved alongside the redesign of port cities where cargo-handling facilities become gradually insulated from urban life, especially through the deployment of technical artifacts for surveillance and security such as video cameras, fences, barbed wire, and security guards.[67]

The port cities located in Chile's major primary-commodity export zones illustrate the increasing imbrication of urban space within complex transnational networks of logistical connectivity. The territorial infrastructures and sociotechnical arrangements that connect large-scale mining in Chile with Asian and European markets are astonishing in their scale, operational dynamism, and technological sophistication. Chile boasts twenty-four multipurpose ports and thirty private terminals specializing in minerals and oil.[68] Mining activity is largely concentrated around the Atacama Desert. Despite being the driest desert in the world, it is traversed by constant flows of commodities, energy, and people. A wide array of pipelines, railways, highways, road networks, transmission lines, and dry-bulk carriers gravitate around the arid landscapes of this infrastructural corridor. Besides mere transport of minerals, commuting networks have also expanded dramatically as the mining industry has become increasingly capital-intensive and in need of an ever more qualified workforce. An official

---

One Road Initiative, thesis, Harvard University, 2017; Seth Schindler and Juan Miguel Kanai, "Peri-Urban Promises of Connectivity: Linking Project-Led Polycentrism to the Infrastructure Scramble," *Environment and Planning A*, March 11, 2018.

67   Alberto Toscano, "Lineaments of the Logistical State," *Viewpoint*, September 28, 2014, viewpointmag.com; Cowen, *Deadly Life of Logistics*; Jorge Budrovich Sáez and Hernán Cuevas Valenzuela, "Contested Logistics? Neoliberal Modernization and Resistance in the Port of Valparaíso," in Jake Alimahomed-Wilson and Immanuel Ness (eds.), *Choke Points: Logistics Workers Disrupting the Global Supply Chain* (London: Pluto Press, 2018).

68   Ministry of Public Works, *Infraestructura portuaria*.

of Codelco recounts that the Calama airport went from servicing two flights to Santiago a week in 2001 to seventeen per *day* in 2016.[69] The main mining towns of the whole biogeographical region structured around the Atacama Desert are Vallenar, Copiapó, Tocopilla, Antofagasta, Calama, and Iquique.

Antofagasta is where the major logistical networks of the mining industry converge. The city's four port complexes—Angamos, Antofagasta, Coloso, and Mejillones—constitute the centerpiece of a clockwork mechanism that connects the sea to the vast mineral deposits that are constantly wrested from the surrounding mountains. Together, these ports handled 11.45 million tons of cargo in 2011, making them the largest port complex in all of Chile, with 18 percent of the country's total port volumes.[70] All of these facilities are almost exclusively dedicated to the mining industry, with the main products handled being copper concentrate and copper cathodes. These ports have grown dramatically, doubling their throughput in just over a decade. Volumes handled went from 6.4 million tons in 2003 to 11.4 million tons in 2013.[71] An estimated 100 trucks and 200 train wagons entered Antofagasta on a daily basis in 2013.[72] Because the ports of Mejillones and Angamos are located in "greenfield sites" (outside densely populated areas), they handle a substantially higher volume of ground transport than Antofagasta's ports.[73]

Walking through the streets of Antofagasta is disorienting. More than an urban agglomeration, the city feels like an autonomous mechanical apparatus that constantly pumps the flows of extraction (minerals, fixed capital, and living labor) into a giant circulatory system. Whereas the infrastructures that mediate the metabolism of cities are usually encased within walls and buried underground so that they remain occluded to everyday urban experience, Antofagasta's infrastructures are simultaneously their built environment; the city *is* the infrastructure. Rather than being designed to mediate encounters—as Henri Lefebvre said of cities in general—Antofagasta's urban geography appears designed to mediate *performance*. The frenzied

---

69   Interview with a Chilean mining executive, January 10, 2017.
70   OECD, *Competitiveness of Global Port Cities*, 5.
71   Ibid., 12.
72   Ibid., 17.
73   Interview with a Chilean port company executive, January 27, 2017.

movement of port cranes, cargo ships, trains, trucks, and industrial workers gives the impression that the sprawling technological systems that converge in the city are beyond any possibility for human control. Deborah Cowen suggests that the design and production of logistical cities marks a new phase in the production of urban infrastructure specifically tailored to underpin the movement of global supply chains.[74] However, the imperative to protect such flows from potential obstructions, she argues, has also marked the emergence of new forms of militarization and political-geographical enclosure in logistics cities across the globe.

That sense of a hostile, autonomous power ruling the life of the city has become an important source of social discontent in Antofagasta. To begin with, the aggressive push toward port automation and technological modernization that began with Law 19.542 has had devastating effects upon local labor markets. Besides mass layoffs, new technologies for cargo handling have also translated into rampant labor casualization: an estimated 70 percent of Chile's port workers now work under temporary contracts.[75] With Law 20.773 in 2014, living labor became even more degraded, as port operators were granted permission to hire workers with eight-hour work contracts, often renewed daily.[76] A leader of a port workers' trade union recalls that every time a major automatic crane, operating system, haulage device, or any other automation technology was implemented at the ports of Valparaíso (one of the main port cities in Chile), some of his coworkers were pushed into abject poverty and homelessness by the concomitant layoffs. He also noted that it has been emotionally devastating to see some of his former colleagues turned into beggars and panhandlers.[77]

The staggering income inequality that has resulted from recent concentration and capitalization in the port and mining industries has also exacerbated rent gaps in Antofagasta. Those who are not employed by the mining or port industries can hardly afford to pay the rent and are often displaced to an expanding constellation of *campamentos* in the

---

74   Cowen, *Deadly Life of Logistics*, 173.
75   Fundación Chile, *Trabajadores portuarios*, 22.
76   Ibid., 22.
77   Interview with a leader of a nationwide port workers' union in Valparaíso, January 18, 2017.

outskirts of the city. As noted in the previous chapter, the number of families living in *campamentos* in Antofagasta skyrocketed from 632 in 2007 to 6,229 in 2016.[78] Furthermore, as a result of mineral dust being constantly released into the atmosphere by trains and trucks carrying copper products as well as water pollution from port operations, public health has deteriorated substantially. Blood and urine samples collected in 2015 confirmed the presence of nineteen heavy metals in Antofagasta's atmosphere and water.[79] In this sense, the seamless motion of the logistical systems of the mining supply chain is starkly juxtaposed with the sclerotic spaces of impoverishment, environmental degredation, and social suffering that surround them. Diverse political initiatives have sought to galvanize local constituencies into action to tackle these problems. The general sense, however, is that the reorganization of logistical space (the real cause of the problems) is simply impervious to local democratic control.

Since the deployment of infrastructures for connectivity severely disrupts local environments and livelihoods, Chile's legal frameworks for territorial planning and design express a sense of liberal emergency against the tacit class enemy that informs much of late liberal thought and institutional practice. The decision-making process is overwhelmingly concentrated at the national level, and local governments have retained very few—if any—prerogatives. In a report on Antofagasta's ports, the OECD states that port infrastructure does not appear prominently in city or regional plans.[80] This, the report concludes, "is a function of the way ports are governed in Chile, heavily influenced by the private sector and central government, with little involvement at the regional or local level." As a result, the OECD notes, ports in Chile tend to develop independently of the cities in which they are located. Such an institutional design is not deliberate but a response to actual perils in the smooth functioning of logistical networks. In a 2015 report, ECLAC highlights labor insurgency in the port industry as one of the main challenges to the governance of supply chains. From 2010 to 2014, ECLAC counted

78   Techo Para Chile, "Catastro de Campamentos 2016," 60.
79   La Tercera 2015, "Masiva marcha contra la contaminación en Antofagasta," diario.latercera.com, accessed April 16, 2017.
80   OECD, *Competitiveness of Global Port Cities*, 21.

312 days of strikes.[81] As a consequence, it recommended that governments rebalance the existing relationship between centralization and decentralization, "leaning towards the return to centralization, especially with respect to decision-making processes related to territorial design and planning."[82]

The APEC also advocates severing port planning policy even further from effective democratic control. In a 2016 report, it argues that Chile should consider establishing an independent port regulator (one that is not directly appointed by government officials). The advantage, according to the APEC, is that this "increases transparency and potentially moves regulation one step away from the political process."[83] Unsurprisingly, these institutional reconfigurations often manifest in everyday experience in Antofagasta and other northern port cities as trenchant forms of corporate and state violence. Lack of democratic accountability and the intervention of police forces when mineral flows are disrupted by social protest are the main ways the alien objectivity of the state as the fetishized expression of the "logistics turn" becomes palpable for the laboring classes. Such contexts are where Holloway considers that the apparent neutrality and fragmentation of the forms, their mystifying disconnections, tend to come into constant conflict with workers' total experience of class oppression.[84] For organized labor in the port and mining industries, the logistical infrastructures of the mining supply chain and the territoriality of state power are not two different things. Strikes in these industries have been increasingly taking the form of blockades and technical sabotage, tactics understood to be at once labor insurgency and political protest against the state.

Far from being specific to resource extraction, blockages and technical disruptions have become common tactics for the logistical sector as a whole. Just as sabotaging systems of machinery was a popular tactic for organized labor in previous stages of capitalist development, infrastructure blockades have become social movements' staple struggle tactic at logistical "chokepoints." To physically attack the flows

---

81   ECLAC, *El gran desafío*, 86.
82   Ibid., 99.
83   APEC, *Case Study on the Role of Services Trade*, 10–11.
84   Holloway, "State and Everyday Struggle."

facilitated by logistical infrastructures, the Invisible Committee claims, "is to politically attack the system as a whole."[85] Perhaps one of the most subversive elements of the blockade is not merely that it disrupts the ebb and flow of the circuits of capital, but that it anchors and actualizes the philosophical categories of class antagonism into the lived experience of workers and communities. According to Arturo Giovanitti, a labor organizer for the 1912 Lawrence textile strike in the United States, "it is only when [worker sabotage] becomes an idea that it becomes a dynamic and disintegrating force of bourgeois society."[86] Sabotage wrests from the political state one of its cardinal faculties: organizing and regulating the forces of production. In so doing, it transfers these faculties to workers.[87] In the face of rampant labor precariousness and deteriorating public health, Antofagasta port workers initiated a cycle of strikes in 2013 that brought the whole logistical apparatus of extraction to an abrupt halt. Workers at the ports of Angamos and Mejillones were demanding half an hour of remunerated lunch time and protesting the precarious conditions of those hired under eight-hour contracts. The strike quickly spread to the vast majority of ports across Chile. Industrial actions included blocking roads, setting up barricades, and suspending all port operations.

The state was swift and uncompromising in its response. To break the strikes of Angamos and Mejillones, the police and other armed forces intervened with forty antiriot vehicles armed with water and tear-gas cannons, fifteen police trucks, and more than sixty other vehicles, as well as large contingents of police in antiriot gear and armed navy officials.[88] According to a local newspaper, many of the workers were injured as a result of unsparing police brutality. In addition, the police allegedly took some of the picketing workers to another place, where they were beaten and even tortured.[89] These forms of police repression in spaces of extraction have become standard practice in Chile and throughout Latin

---

85   Invisible Committee, *To Our Friends*, 93.

86   Cited in Rebecca Lossin, "Capitalist Saboteurs," *Jacobin*, 2016, jacobinmag.com.

87   Lossin, "Capitalist Saboteurs."

88   Antonio Justo, "Chile: ¡Viva la huelga portuaria!" Partido de los Trabajadores Socialistas, January 9, 2013, pts.org.ar.

89   *Diario Antofagasta*, "Violenta represión policial contra portuarios de Mejillones en paro," March 19, 2013, diarioantofagasta.cl.

America. In 2012, a similar incident had taken place in the port town of Huasco, on the southern end of the Atacama Desert. Local communities affected by mineral extraction and thermoelectric power generation decided to block the roads that connected the ports of Huasco to the mines and faced brutal police repression as a result.[90] The militarization of the supply chain in Chile and Latin America has not been a haphazard process, but is underpinned by standards of "best practices" endorsed by multilateral organizations.[91]

As a port workers' union member claims, strengthening supply-chain security has not merely escalated physical state repression. It has also involved systematic intimidations, threats, and harassment by domestic intelligence agencies. Kafkaesque incidents in which secret police harass union members by following them in cars at night or detaining them for brief periods without disclosing criminal charges, the interviewee argues, have become standard state practice.[92] Although Chile's Anti-Terrorism Statute was originally passed by the parliament through Law 18.314 of 1984, subsequent amendments have broadened the powers and functions of the organs of surveillance and supervision to prevent behaviors deemed by the existing legislation to instill fear in the civilian population. In March 2018, President Sebastián Piñera proposed a new amendment to Law 18.314 to introduce new surveillance and investigation techniques, such as undercover agents, interception of communications, and informants.[93] This proposed amendment aroused controversy among the public opinion and multilateral and civil-society organizations, which argued that it would lead to new ways to repress and criminalize sociopolitical and labor protest.[94]

---

90   Arboleda, "In the Nature of the Non-City."

91   World Bank, *Supply Chain Security Guide* (Washington, D.C: World Bank, 2009), documents.worldbank.org/curated/en/862601468339908874/pdf/579700 WP0SCS1G10Box353787B01PUBLIC1.pdf; ECLAC, *Seguridad de la cadena logística*.

92   Interview with a leader of a nationwide port workers' union in Valparaíso, January 18, 2017.

93   *El Mercurio*, "Los 11 cambios a la ley antiterrorista que impulsará el Gobierno de Piñera," March 23, 2018, emol.com/noticias/Nacional/2018/03/23/899863/ Los-10–cambios-a-la-Ley-Antiterrorista-que-impulsara-el-Gobierno-de-Pinera. html.

94   *El Desconcierto*, "Repetir lo que no funciona: 6 riesgos en los que incurre la

Seen in this light, the evolving formations of state power that underlie the emergence of logistical urbanization insulate the decision-making process from democratic control and intensify the use of organized state violence to dissipate dissent and social mobilization, especially if it threatens to disturb the homeostasis of logistical systems. The liberal state—as theorized by authors in the German ordoliberalism tradition—employs the "power of the commonwealth" to enforce the practice of justice, according to Bonefeld.[95] The state, from this perspective, is responsible for securing the proper use of freedom—by means of police. Its task, Bonefeld explains, is to punish the misuse of freedom and thus secure the law of private property. This is why the historical composition of the state during fascism cannot be seen as an "exceptional" form of state, as Poulantzas would suggest.[96] Rather, "the coercive character of the state exists as presupposition, premise and result of the social reproduction of class

---

esperada reforma antiterrorista de Sebastián Piñera," March 28, 2018, http://www. eldesconcierto.cl/2018/03/28/repetir-lo-que-no-funciona-6-riesgos-en-los-que-incurre-la-esperada-reforma-antiterrorista-de-sebastian-pinera/.

95   Bonefeld, *Strong State*, 171–72. For the German ordoliberal tradition, according to Bonefeld, the relationship between economy and state is an innate one. According to their proponents, the economy has no independent existence; rather, its independence amounts to a *political event*, one that needs to be reasserted again and again in order to prevent the illiberal use of freedom. The foundational statements of ordoliberal thought show keen understanding of Schmitt's political theology, "ranging from a critique of mass democracy . . . from calls for a political decision for a commissarial dictatorship to the use of language and phraseology." For ordoliberal authors, organized coercive force is required so that the freedom to compete does not "degenerate into a vulgar brawl" that threatens to undermine it (Röpke, cited in Bonefeld, *Strong State*, 24). Ordoliberalism thus develops the necessity of the state as the authoritative force binding the process of social reproduction at the system-wide level. Although ordoliberalism may appear to be a subterranean intellectual tradition with no bearing on the forms of neoliberal economic thought popularized by the Chicago School years later—and which informed Latin America's economic shift toward market liberalization after the 1970s—Bonefeld shows that in fact the differences between the two are of nuance, not of doctrine. It was actually Alexander Rüstow—one of the central theoreticians of ordoliberalism—who coined the term neoliberalism in 1938 to distinguish the "new liberalism" from the tradition of laissez-faire liberalism (Bonefeld, *Strong State*, 10). Rüstow developed this term in sharp opposition to Ludwig von Mises, whom he considered to be a "paleo-liberal" with blind faith in the "natural" capacity of markets to self-regulate.

96   Referenced in Bonefeld, "Social Constitution," 120.

antagonism ... and not as a qualitative new period of capitalist development."[97]

Conclusion

The logistics turn in the extractive industries has underpinned a process of technological-organizational restructuring whereby systems of transport and circulation—hitherto considered exclusively in terms of cost—have been recast as internal elements in producing economic value. This organizational shift has been contingent on the design and production of built and unbuilt environments that are functional to a modality of primary-commodity production structured around speed, homeostasis, and flow. In Chile, logistical spaces of extraction have been fractured right through the middle—between the seamless, escalating velocity of mineral flows crisscrossing the Atacama Desert and the sclerotic fabrics of urbanization that coexist with them. Recent port expansion and automation undertaken to cope with increasing levels of mineral output have brought with them labor casualization, trenchant income inequality, and aggressive policing for those in labor unions and socioenvironmental movements. Constant dry-bulk carrier, train, and truck traffic has also led to severe air, water, and noise pollution in mining towns. Logistical and infrastructural hubs of the mining industry, infamously dubbed "*zonas de sacrificio*" ("sacrifice areas"), have seen cancer epidemics, a wide array of respiratory diseases, and declining yields in subsistence agriculture. Far from being specific to Chile, the technologies, policy frameworks, and planning instruments that give momentum to logistical urbanization are being reproduced in commodity-exporting zones across the global economy.[98]

In broad terms, logistical urbanization entails the increasing imbrication of urban space within the sociotechnical infrastructures of circulation that animate world trade, as well as the military and disciplinary

97   Bonefeld, "Social Constitution," 120.
98   Cowen, *Deadly Life of Logistics*, chapter 5; Toscano, "Lineaments of the Logistical State"; Khanna, *Connectography*; Schindler and Kanai, "Peri-Urban Promises of Connectivity."

apparatuses of the capitalist state. The astonishing reconfiguration and upgrading of industrial technology that has enabled Asian economies to cast thick logistical webs through entire oceans and continents and gain access to new resource frontiers is leading to the formation of new megaregions. In China's One Belt, One Road initiative, for example, connectivity infrastructure—in the form of high-speed rail networks, maritime trade routes, port cities, and highways—is designed to recast Eurasia as a single "megacontinent."[99] Although this initiative and the "new silk roads" might be the most paradigmatic examples of this trend, a wide array of macroregional planning frameworks and logistics corridors for the design of export-led economies in other parts of the world have been attracting scholarly attention in recent years.[100]

This logistics turn is giving rise to important transformations in the political authority of the modern state, especially toward configurations that combine—in contradictory yet complementary ways—the internationalization and concentration of state power and its institutional arrangements. These emerging formations of liberal statecraft cast into stark light the fact that the accumulation of capital is global in content but national in form.[101] The contradictory yet unitary logic of state power and capital accumulation is materially embodied in the police trucks, water cannons, and tear-gas canisters unleashed on picketing port and mine workers in the industries across Latin America. Such are the political forms assumed by the global unfolding of value. The political authority of the latter, the chapter has illustrated, continues to be mediated at the national scale. For this reason, preserving the

99   Khanna, *Connectography*; Weidong Liu and Michael Dunford, "Inclusive Globalization: Unpacking China's Belt and Road Initiative," *Area Development and Policy* 1, no. 3 (2016): 323–40. For the relationship between Latin America and the Belt and Road Initiative, see Margaret Myers, "China's Belt and Road Initiative: What Role for Latin America?" *Journal of Latin American Geography* 17, no. 2 (2018): 239–43.

100   Japhy Wilson, "Colonising Space: The New Economic Geography in Theory and Practice," *New Political Economy* 16, no. 3 (2011): 373–97; Juan Miguel Kanai, "The Pervasiveness of Neoliberal Territorial Design: Cross-Border Infrastructure Planning in South America since the Introduction of IIRSA," *Geoforum* 69 (2016): 160–70; Simón Uribe, "Illegible Infrastructures: Road Building and the Making of State-Spaces in the Colombian Amazon," *Environment and Planning D: Society and Space*, August 6, 2018; Neil Brenner, *New Urban Spaces: Urban Theory and the Scale Question* (Oxford: Oxford University Press, 2019); Kanai and Schindler, "Peri-Urban Promises of Connectivity."

101   Charnock and Starosta (eds.), *New International Division of Labour*.

homeostasis of logistical extraction networks has become a matter of state security, even in so-called postneoliberal governments such as those of Bolivia, Ecuador, and Venezuela. In the next chapter, I continue to interrogate the state apparatus to understand the role of technical expertise in producing spaces of extraction.

5. EXPERTISE

Technocracy and Expropriation

Economists were present at the creation of the cyborg sciences, and, as one would expect, the cyborg sciences have returned the favor by serving in turn to remake the economic orthodoxy in its own image.[1]

The law of value contains the force of law-making violence within its concept—in its civilized form, it appears as the freedom of economic compulsion.[2]

Introduction

Setting the interconnected infrastructures of extraction into motion is no trivial matter, and by no means is it exclusively contingent upon the deployment of fixed and financial capitals. Modern science, and especially the algorithmic and epistemological systems of neoclassical economics mobilized by technical experts in state agencies, is also one of the key driving forces of the spatial technologies, population flows, and land-tenure schemes that act as the foundation of the planetary mine. Generally speaking, sensuous social praxis could not be

---

1 Philip Mirowski, *Machine Dreams: Economics Becomes a Cyborg Science* (Cambridge: Cambridge University Press, 2002), 6.
2 Bonefeld, *Critical Theory*, 82.

remodeled into economic categories recast as if they were "laws of nature" without the categories of thought and epistemological systems of "economic science," as well as its most emblematic incarnation: the technocrat. This chapter offers an account of how the environment-making process that transformed Chile into one of Latin America's main exporters of raw materials manifested in the inverted form of seemingly timeless economic abstractions—economic ideas of price signals, incentives, efficiency, equilibria, and so forth. The complex and specialized character of these economic categories, I argue, obfuscates an underlying political project driven by very specific class interests.

Comprehensive reforms in terms of water and land rights, electricity production, and mining implemented from the mid-1970s onward and inspired by neoclassical economic principles cast into relief the fact that capitalism is an inherently geographical project whose central mediating mechanism is the modern state.[3] This chapter argues that the environment-making powers of the state are, as authors in the form-analysis tradition show, a fetishized manifestation of the class antagonism of modern society.[4] Notably, one of the main outcomes of implementing policy frameworks inspired by monetarist economic thought—and other related neoclassical approaches to economic theory—has been the creation of a whole new gamut of property owners and rentiers in Chile. Creating and enforcing new property rights, this chapter shows, has been directly contingent on separating small producers from their means of subsistence and on systematically privatizing common goods. I build upon recent interventions that propose to understand primitive accumulation as an *ongoing* mode of social labor premised on a logic of violent expropriation, the central agent of which is the state.[5]

3   Christian Parenti, "Environment-Making in the Capitalocene: Political Ecology of the State," in Jason W. Moore (ed.), *Anthropocene or Capitalocene? Nature, History, and the Crisis of Capitalism* (Oakland, CA: PM Press, 2016).

4   Bonefeld, *Critical Theory*, chapters 7 and 8; Simon Clarke, "State, Class Struggle, and the Reproduction of Capital."

5   Werner Bonefeld, "Primitive Accumulation and Capitalist Accumulation: Notes on Social Constitution and Expropriation," *Science and Society* 75, no. 3 (2011): 379–99; Bonefeld, *Critical Theory*, chapter 4; Parenti, "Environment-Making in the Capitalocene"; William Clare Roberts, *Marx's Inferno: The Political Theory of Capital* (Princeton, NJ: Princeton University Press, 2017); William Clare Roberts, "What Was Primitive Accumulation? Reconstructing the Origin of a Critical Concept," *European Journal of Political Theory*, October 11, 2017; Singh, "On Race, Violence, and So-Called Primitive

Contemporary readings of resource extraction often invoke "accumulation by dispossession," "expulsions," and "primitive accumulation," as these notions place force and extortion at the very center of how the primary-commodity sector functions. Although insightful, some of these accounts conflate capital with conquest and usurpation, and therefore implicitly end up offering a reading of capitalism that does not differ in substantial respects from feudalism—where interpersonal coercion and force were the pivotal causal mechanism of socioeconomic life. To elucidate what is historically unique to making spaces of resource extraction in modern society, I draw from alternative readings of the Marxian notion of primitive accumulation that take seriously the persistence of violence and coercion, but within the framework of the indirect and impersonal forms of social mediation characteristic of capitalist modernity.[6] A common thread in these readings is the claim that liberal society sees itself as antithetical to violence but opportunistically harnesses the violent acts carried out by others, especially the state and landlords. To fully flesh out the distinction between the violence of *capitalism* and the opportunism of *capital*, I focus on Chile's neoliberal technocracy. Stories of the Chicago Boys and of monetarist experiments implemented during the Pinochet dictatorship are part and parcel of the foundational myth of neoliberal globalization and have been construed and exported as a template for the design and regulation of export-oriented economies across the world.

Latin America's turbulent history with neoliberal experiments, in the context of military dictatorship and structural-adjustment programs sponsored by multilateral organizations like the International Monetary Fund and World Bank, has made technical expertise a fundamental aspect of governance structures, politics, and policymaking across the board.[7] Yet, far from being an anomalous case, Latin America is a

---

Accumulation"; Diego Andreucci, Melissa García-Lamarca, Jonah Wedekind and Erik Swyngedouw, "Value Grabbing: A Political Ecology of Rent," *Capitalism Nature Socialism* 28 , no. 3 (2017): 28–47.

6   McNally, "Blood of the Commonwealth"; Bonefeld, "Primitive Accumulation and Capitalist Accumulation"; Roberts, *Marx's Inferno*; Roberts, "What Was Primitive Accumulation?"; Ballvé, "Everyday State Formation."

7   Arturo Escobar, *Encountering Development: The Making and Unmaking of the Third World* (Princeton, NJ: Princeton University Press, 1995; Patricio Silva, "Technocrats and Politics in Chile: From the Chicago Boys to the CIEPLAN Monks," *Journal of Latin American Studies* 23, no. 2 (1991): 385–410; Silva, *In the Name of Reason: Technocrats*

microcosm of a global shift away from politics, government, and "dissensus"[8] and toward governance, the rule of expertise, and "value-free" policy formulation since the 1970s.[9] This shift is particularly relevant for explorations of contemporary global sociospatial change; indeed, as Brenner has noted, the operational landscapes of extended urbanization are being comprehensively engineered through large-scale territorial planning strategies and neoliberal governance frameworks in order to link their developmental rhythms to large zones of agglomeration.[10] Governance is crucial to this process because capital can only employ the world-making capacities of nature (human laborers, natural resources, territory) if the state has first seized portions of the earth through extraeconomic force. It is in this sense that territory needs to be viewed above all as a form of *political technology*; that is, a spatial category that is measured, demarcated, bordered, represented, and policed eminently by the lawmaking violence of the state.[11]

To get at the political root of large-scale territorial design, I interrogate further the modalities of liberal authoritarianism discussed in the previous chapter, but center their origins more directly on the categories of neoclassical economic thought that became ensnared in the decision-making fabric of the Chilean state after the military coup of 1973. The state, as previous chapters have highlighted, is the political form of

---

and Politics in Chile (University Park: Pennsylvania State University Press, 2008); Miguel Centeno and Patricio Silva (eds.), *The Politics of Expertise in Latin America* (London: Macmillan, 1998); José Ossandón, "Economistas en la élite: entre tecnopolítica y tecnociencia," in Alfredo Joignant and Pedro Guell (eds.), *Notables, tecnócratas y mandarines: elementos de sociología de las élites en Chile* (Santiago: Ediciones Universidad Diego Portales, 2011).

8   Erik Swyngedouw and Japhy Wilson (eds.), *The Post-Political and Its Discontents: Spaces of Depoliticization, Spectres of Radical Politics* (Edinburgh: Edinburgh University Press, 2015).

9   Timothy Mitchell, *Rule of Experts: Egypt, Techno-Politics, Modernity* (Berkeley: University of California Press, 2002); Carlos De Mattos, "De la planificación a la governance: Implicaciones para la gestión territorial y urbana," *Revista Paranaense de Desenvolvimento* 107 (2004); Paul Cammack, "The Governance of Global Capitalism: A New Materialist Perspective," *Historical Materialism* 11, no. 2 (2003): 37–59.

10   Brenner, "Urban Theory without an Outside."

11   Stuart Elden, "Governmentality, Calculation, Territory," *Environment and Planning D: Society and Space* 25 (2007): 562–80; Stuart Elden, "Land, Terrain, Territory," *Progress in Human Geography* 34, no. 6 (2010): 799–817; Parenti, "Environment-Making in the Capitalocene."

market liberty.[12] It is then symptomatic that Friedrich von Hayek, one of the main architects of the Chicago School of neoclassical economics, considered the Pinochet dictatorship in Chile to be "liberalizing." Specifically, he believed it resolved the "excess of democracy," governing for personal freedom and the free economy.[13] On this basis, the chapter begins by briefly reflecting on the nature of modern science with the aim of understanding why the principles of rational abstraction that underpin mainstream economics cannot be considered in isolation from the dynamics of expulsion and coercion at the heart of the Marxian notion of primitive accumulation. The second section provides a historical overview of the emergence of Chile's neoliberal technocracy and its relationship with state power. The remainder of the chapter brings to life these theoretical and historical insights by examining the role of technical expertise in the design of specific policy apparatuses for governing water, energy, mining, and land. The section shows how such policy frameworks laid out the legal and economic foundations for the infrastructural systems that connect primary-commodity production in the Chilean Andes with manufacturing and construction in Asia.

On Economic Rationality, Expropriation, and the State

According to Saskia Sassen,[14] the complexity of contemporary formations of scientific knowledge and techno-scientific praxis often obfuscates and renders imperceptible their predatory character. For this reason, authors in the form-analysis tradition are adamant that the critique of political economy entails thinking against the spell that dazzling economic forms have cast upon social reality.[15] As Bonefeld points out, the reified world of economic necessity is innately practical: "It entails the actual relations of life in their inverted economic

---

12   Bonefeld, "Adam Smith and Ordoliberalism."

13   Bonefeld, *Strong State*, 62; Karin Fischer, "The Influence of Neoliberals in Chile before, during, and after Pinochet," in Philip Mirowski and Dieter Plehwe (eds.), *The Road from Mont Pelerin: The Making of the Neoliberal Thought Collective* (Cambridge: Harvard University Press, 35).

14   Saskia Sassen, "Predatory Formations Dressed in Wall Street Suits and Algorithmic Math," *Science, Technology and Society* 22, no. 1 (2017): 1–15.

15   Bonefeld, *Critical Theory*.

form."[16] The critique of political economy involves an effort to grasp the human and ecological content of the "social hieroglyphs" that we usually take for objective reality. Academic economics, Backhaus explains, only knows the result of these displaced, alienated forms of thought.[17] The origins of such mystical economic things in pillage, expropriation, and labor exploitation, however, remain sidelined from analysis. In fact, historians of science point out that the principles of rational abstraction at the heart of modern science are premised upon the necessity to objectify the world as a means to exert direct control over the phenomena observed.[18] As a form of knowledge that emerged and gradually evolved in service of another abstraction—commodity exchange—science was inherently ruled by a logic of appropriation.[19]

To the extent that neoclassical economics evolved by mirroring the natural sciences in form and content, it also internalized the logic of

---

16 Ibid., 8.

17 Backhaus, "Between Philosophy and Science: Marxian Social Economy as Critical Theory."

18 Mumford, *Technics and Civilization*; Sohn-Rethel, *Intellectual and Manual Labour*. During late phases of feudalist society, Sohn-Rethel recounts how intellectual labor gradually emancipated itself from manual labor, as artisans did not have access to the logic of socialized thought of mathematics. In particular, the Galilean assumption of inertial motion opened the applicability of mathematics to the calculation of real, physical motion. This founding principle, Sohn-Rethel explains, provided the methodological basis for modern science, as it allowed natural phenomena to be isolated from their environment and tested experimentally. These scientific apparatuses did not develop in a vacuum, as they tended to mirror changes taking place in actual social practice, especially as the value-form of social relations slowly but steadily asserted itself as the dominant mode of social reproduction in the modern world. The contribution of Sohn-Rethel's landmark study consists in demonstrating that the rise of modern science is inwardly connected to the rise of modern capitalism.

19 Hilary Rose, "Hand, Brain and Heart: A Feminist Epistemology for the Natural Sciences," in Harding (ed.), *Feminist Standpoint Theory Reader*; Moore, *Capitalism in the Web of Life*, chapter 8; Donna Haraway, *Simians, Cyborgs and Women: The Reinvention of Nature* (New York: Routledge, 1991). According to Haraway, this feature of modern science became much more pronounced in the twentieth century, when the life sciences moved from physiology to systems theory. An organismic model of socionatural systems, she argues, facilitated the conception of society as a balanced, harmonious whole, therefore making it the ideal raw material of engineering. Engineering thus became the guiding logic of science in the twentieth century and entailed the "rational placement and modification of human raw material in the common interest of organism, family, culture, society, industry" (Haraway, *Simians, Cyborgs and Women*, 48).

appropriation that followed from such elemental principles of radical abstraction.[20] As Mirowski shows, the marginalist revolution in economics underpinned a shared vision of the operation of the market that was avowedly mechanical in a physical sense of the term. Economics became a "science" of "causality, rigid determinism, and preordained order; in other words, it was physics prior to the second law of thermodynamics, a science most assuredly innocent of the intellectual upheavals beginning at the turn of the [twentieth] century."[21] Although abstraction and the fragmentation of knowledge allowed modern science to make nature "legible" for capital accumulation,[22] such features became even more pronounced at the onset of the third machine age. This was when "research and development" became a separate branch within the division of labor of large companies, and state-funded scientific work was reoriented to address the strategic requirements of accumulation.[23] Like any other business, "research" became exclusively concerned with maximizing profit and accelerating turnover times of capital.

Under the technological context of the third machine age, the unrelenting proletarianization of scientific labor fragmented the sciences and, according to Mandel,[24] gave rise to overspecialization and "expert idiocy." For Mandel, the university ceased to be concerned with the production of educated people who could grasp the imbrication of specialized knowledge with society, polity, and economy; instead, it had become invested in producing intellectually skilled wage-laborers to fit the needs of late-capitalist technology. Unsurprisingly, the disciplinary context of economic science mirrored the types of expertise and scientific labor that began to emerge in the natural sciences as a whole. For Fine and Milonakis, the process of formalization of "economic science" has led to a whole generation of scientists with excellent technical capacities but who are incompetent to understand the functioning of the

20   Simon Clarke, *Marx, Marginalism and Modern Sociology* (London: MacMillan, 1991 [1982]); Milonakis and Fine, *From Political Economy to Economics*; Mirowski, *Machine Dreams*.

21   Mirowski, *Machine Dreams*, 7.

22   Moore, *Capitalism in the Web of Life*, 199.

23   Mandel, *Late Capitalism*; Tony Smith, "Red Innovation," *Jacobin* 17 (2015): 75–82.

24   Mandel, *Late Capitalism*, 263.

economy—their actual object of study.[25] The disciplinary configuration of "economic science" has tended to mirror the ontological complexion of automated systems, Mirowski explains, because Walrasian economics has been from the very outset fascinated by the principles of natural order embedded in the machine.[26] Machine rationality and machine regularity, in Mirowski's view, are the constants in the history of neoclassical economics. In fact, William Stanley Jevons, one of the foremost architects of the marginalist revolution, proudly compared the proverbial "rational agent" of economic models to a machine.[27]

The active appropriation of the elements of abstraction and regularity that are intrinsic to the process of mechanization explains why economic models often assume the form of alienated, self-acting forces over the individuals that produce them. It also explains why neoclassical economics has been so enamored of strong state authority. Mitchell argues that the possibility for "economics" to reinvent itself as the set of mathematized standards of representation for observing the "economy"—understood as an object or machine of sorts—belongs to the history of empire and the collapse of the colonial order.[28] In particular, Mitchell illustrates how the India Office in London served as a seedbed for a new breed of economists—including John Maynard Keynes—who analyzed Indian currency and finance at a distance, as detached observers, and in so doing drew the blueprints for the key assumptions of contemporary economic science. In Mitchell's view, modern colonial rule thus opened a space of separation and a relationship of curiosity that made it possible for economists to construe a set of flows and relations as a "case," a "self-contained object whose 'problems' could be measured, analyzed and

---

25 Ben Fine and Dimitris Milonakis, " 'Useless but True': Economic Crisis and the Peculiarities of Economic Science," *Historical Materialism* 19 , no. 2 (2011): 3–31.

26 Mirowski, *Machine Dreams*.

27 Ibid., 9.

28 Mitchell, *Rule of Experts*, 4. Mitchell's study about the genesis of neoclassical economics demonstrates how the notion of "the economy" in its contemporary sense did not appear until the mid-twentieth century, when economists formulated the concept to mean "the totality of monetarized exchanges within a defined space." The economy thus came into being as a "self-contained, internally dynamic, and statistically measurable sphere of social action, scientific analysis and political regulation." Before this, *economy* carried the older meaning of "thrift," which referred particularly to the rational management and utilization of resources, a notion that was expanded to the level of the political order by the classical political economists.

addressed by a form of knowledge that appears to stand outside the object and grasp it in its entirety."[29]

As this chapter will demonstrate, monetarist experiments with the Chilean economy in the context of the Pinochet dictatorship resonate in manifold ways with the colonial and authoritarian basis that Mitchell attributes to neoclassical economics. For Milton Friedman and his University of Chicago colleagues (known as the "Chicago Boys"), 1970s Chile offered an ideal opportunity to "stress-test" ideas about incentives, the rationality of market actors, price controls, and monetary supply within the domain of a "real-life" national economy. Against the background of US diplomatic and military interventionism resulting from the Cold War, Chile offered the separation required for detached analysis and subsequent formulation of policy mechanisms. This separation was not only cultural and geographical but because the leverage required to implement reforms was ensured by state violence. In fact, Bonefeld has argued that Friedman had outspoken sympathies for the modes of liberal authoritarian rule favored by ordoliberal authors. As Bonefeld notes, Friedman decried political liberalism as "a soft, self-defeating approach to mass pressures for income redistribution and employment guarantees, arguing that it presents an 'internal threat'" that arises from attempts at political reform.[30] German Ordoliberalism does not consider "the economy" an independent reality. For authors in this tradition, the "invisible hand" is a practice of government.[31] The ordoliberal state is a strong state, one that is suspicious of mass democracy—under the assumption that it leads to tyranny—and justifies the establishment of a commissarial dictatorship as a means of guaranteeing a free economy.[32]

The ordoliberal approach, Bonefeld notes, recognizes that the political equality of laborers is a danger to the laws of private property, and therefore demands a strong state that does not yield to welfare-seeking trade unions and other labor organizations.[33] Hayek's appraisal of the Pinochet dictatorship in Chile as one that might be "more liberal in its

---

29 Mitchell, *Rule of Experts*, 100.
30 Bonefeld, *Strong State*, 48.
31 Ibid.
32 Ibid.
33 Ibid.

policies than an unlimited democratic assembly"[34] attests to the authoritarian-liberal underpinnings of the type of economic theory developed by the Chicago, Freiburg, and Austrian Schools. Regardless of the internal nuances and differentiations between the intellectual traditions of neoclassical economics, Ludwig von Mises—one of the main Austrian School theoreticians—defines their overarching tenet as follows: "The program of liberalism, summed up in a single word, should read: *property* . . . that is, private ownership of the means of production. All the other demands of liberalism derive from this fundamental demand."[35] It is therefore no coincidence that the increasing influence of neoliberal economists upon the decision-making apparatus of the state has galvanized renewed interest in the notion of primitive accumulation—hastily defined by Marx as "the process which divorces the worker from the ownership of the conditions of his own labour."[36] Although seemingly unrelated and even antithetical, violent expropriation and neoclassical economics complement each other in important ways.

Making sense of how neoliberal technocracy plays into the broader question of primitive accumulation, however, requires a departure from most contemporary interpretations of this notion. Despite some disagreements, the new readings of primitive accumulation have two key features in common. First, they challenge the idea that primitive accumulation is part of a historical past, and propose to understand it as an *ongoing* dynamic within the broader reproduction of capitalist society.[37]

---

34 Cited in ibid.

35 Cited in Werner Bonefeld, "Stateless Money and State Power: Europe as Ordoliberal *Ordnungsgefüge*," *History of Economic Thought and Policy* (in press), 1.

36 Marx, *Capital*, Vol. 1, 875. Some of the most influential approaches of the "new reading" of primitive accumulation are those of Harvey, *New Imperialism*, chapter 4; Jim Glassman, "Primitive Accumulation, Accumulation by Dispossession, Accumulation by Extraeconomic Means," *Progress in Human Geography* 30, no. 5 (2011): 608–25; Silvia Federici, *Caliban and the Witch: Women, the Body, and Primitive Accumulation* (New York: Autonomedia, 2004); and Massimo De Angelis, "Marx and Primitive Accumulation: The Continuous Character of Capital's Enclosures," *Commoner* 2 (2001): 1–22.

37 This point has aroused important controversy among different interpretations, given the style and rationale that undergirds part 8 of volume 1 of *Capital*, where Marx introduces the notion of primitive accumulation. The eight chapters that comprise this section are deliberately "historical" and narrative in style and composition, thereby disrupting the pattern of the theoretically oriented approach that Marx unfolds throughout the previous sections. On this basis, the group of authors associated with the "New Dialectics" tradition constitutes an important counterpoint to the new reading of

Second, they challenge the idea that primitive accumulation arises *exclusively* as capital expands to appropriate its outsides—geographical or otherwise. By contrast, the new reading locates primitive accumulation within the very marrow of the capitalist system.[38] In essence, primitive accumulation exists because, according to Marx,[39] the capital relation presupposes workers' complete separation from any means that would allow them to be self-sufficient. The process that creates the capital relation, argues Marx, can be nothing other than the process of expropriation that effects such separation. Far from a "civilized" process, Marx famously argues, such acts of expropriation are so directly premised on violence, deceit, and terror that their history "is written in the annals of mankind in letters of blood and fire."[40]

Although historical in appearance, the reproduction of the law of value is presupposed on permanent extraeconomic force. The rounds of privatization, deregulation, and commodification of extrahuman natures described in the following sections, therefore, are likewise written in the annals of Chile in letters of blood and fire. The estimated toll of Chile's neoliberal revolution ascends to 3,197 assassinations or "forced disappearances"; 450,000 individuals in exile; and unknown amounts of mass torture, incarceration, and intimidation of real or perceived political opponents.[41]

---

primitive accumulation. For these authors, a value-theoretical interpretation reveals that part 8 is a purely historical digression that cannot but confuse the immanent logic of capital already fleshed out in the first twenty-five chapters of *Capital* (Tony Smith, *The Logic of Marx's Capital: Replies to Hegelian Criticisms* (Albany: State University of New York Press, 1990); Christopher Arthur, *The New Dialectic and Marx's Capital* (Leiden: Brill, 2002); Michael Heinrich, *An Introduction to the Three Volumes of Marx's Capital* (New York: Monthly Review Press, 2012). The category of wage labor is not founded on coercion but on the logical principle of abstract social labor. In other words, once value relations are established, the buying and selling of labor-power takes place between formally equal individuals and the necessity of expropriation and interpersonal coercion vanishes.

38   David Harvey's notion of "accumulation by dispossession" is particularly noteworthy in this regard. The idea that primitive accumulation is manifested in the expansion of capital toward its noncapitalist outsides was fully elucidated by Rosa Luxemburg in her pathbreaking 1913 work, *The Accumulation of Capital*.

39   Marx, *Capital*, Vol. 1.

40   Ibid., 875.

41   Carlos Huneeus, *El Régimen de Pinochet* (Santiago: Editorial Sudamericana, 2000); Stephan Ruderer, "Cruzada Contra el Comunismo: Tradición, Familia y Propiedad en Chile y Argentina," *Sociedad y Religión* 22, no. 38 (2012).

New readings of primitive accumulation, although insightful regarding the permanent presence of violence in liberal society (deemed by liberal thought to be the domain of legally equal and formally free individuals), fail to do the important analytical work of disaggregating the agencies at stake in the act of expropriation. As such, they generally tend to conflate capital with violence, thereby overlooking the historical specificity of the reified forms of social mediation that the previous chapters have been bringing to the forefront. According to William Clare Roberts, the new readings of primitive accumulation are reminiscent of the varieties of Saint-Simonian, Owenite, or Proudhonist socialist thought that considered capitalism a simple continuation of feudalism; that is, a regime of force where the "knights of industry" had merely supplanted the "knights of the sword" and the exploitation of labor was therefore just another manifestation of feudal extortion.[42] Saint-Simonian socialism, Roberts points out,[43] made accumulation by exploitation a moral problem of rentiers accruing and abusing their power over the propertyless. By sidelining what was historically specific to the capital relation (i.e., the extraction of labor-power via impersonal compulsions, not direct forced labor), these approaches were deeply misleading in their moralizing intent. This conflation of capital with violence—or direct interpersonal coercion—is markedly present in contemporary readings of primitive accumulation. According to Harvey, for example, accumulation based upon predation, fraud, and violence is constitutive of the operations of capital.[44] Likewise, Federici claims that Marx was "deeply mistaken" when "he assumed that the violence that had presided over the earliest phases of capitalist expansion would recede with the maturing of capitalist relations."[45]

Marx's notion of primitive accumulation, Roberts notes, challenged the moralist reading of Saint-Simonian thought by showing that within capitalism, "capital is the agent of accumulation by exploitation, not the agent of primitive accumulation."[46] In other words, Marx intended to root the theory of primitive accumulation in the extraeconomic (coercive) mediations of landlords and states. The point is not that "capital

42   Roberts, *Marx's Inferno.*
43   Roberts, "What Was Primitive Accumulation?"
44   Harvey, *New Imperialism,* 144.
45   Cited in Roberts, *Marx's Inferno,* 196.
46   Roberts, *Marx's Inferno.*

has its origins in acts of violence and theft, but that capital has its origin in the opportunistic exploitation of the new forms of freedom created by acts of violence and theft."[47] In other words, capital itself rejects the internalization of violence to its mode of operation; it decries coercion as barbarous, premodern, and illiberal. However, it permanently relies on external agents that perform the acts of expropriation and extortion required for its conditions of existence. This reading of primitive accumulation, Roberts points out,[48] underscores the complementarity of capitalist production and state action but also reveals the irreducible difference between them. In the case of Chile, it explains the bizarre state of affairs whereby business conglomerates and the neoliberal economists associated with the Chicago School and the Mont Pelerin Society saw themselves as the advocates and even precursors of liberty, while turning a blind eye to the political violence unleashed by the military regime after 1973. The 1981 congress of the Mont Pèlerin Society in Viña del Mar, Chile, is a case in point. Despite being organized in a setting of ruthless coercion, policing, and even genocidal violence, this event brought together pundits, economists, and academics from around the world in order to philosophize on the question of liberty.[49]

In general terms, neoliberal economists under the military regime were so enticed by the *problématique* of freedom (construed as noncoercion) that the Chilean constitution of 1980 is titled after Hayek's 1960 book *The Constitution of Liberty*.[50] It is particularly telling that to protect individual freedom and market liberty, this constitution "stipulates the need of strong, centralized state authority that can act as guarantor of the established rule of law."[51] According to Fischer, Chile's 1980 constitution strongly advocated a continuous state of exception that could thwart any opposition to the neoliberal reforms. Here is where the question of expropriation comes fully into being. Primitive accumulation is, above all, eminently premised upon the design, enactment, and implementation of property regimes. That is, the process of primitive accumulation not only provides the material conditions for the exploitation

---

47   Ibid., 207.
48   Roberts, "What Was Primitive Accumulation?"
49   Fischer, "Influence of Neoliberals."
50   Ibid.
51   Ibid., 34.

of labor-power via proletarianization but overhauls the very materiality of the capitalist class by creating renewed varieties of property owners, landlords, and rentiers. Recent readings of rent theory have explicitly foregrounded the mediating role of property regimes and entitlements created by the state, *qua* "ultimate landlord," in contemporary expulsion and expropriation.[52] Ultimately, Parenti argues, it is the "geopower matrix of state-centric, Earth-focused techno-rational practices that open nonhuman nature to effective capitalist exploitation."[53] This contradictory configuration of lawmaking violence, property, and rational calculation, subsequent sections show, has been central to the iterations of large-scale territorial restructuring that have enabled spatial technologies of extraction to come into their own. The next section expounds further on this contradiction by looking in more detail at the nature of technocratic expertise in Chile.

The Rise of Neoliberal Technocratic Rule in Chile

If we consider technocratic experts to be personnel who use their claim to privileged knowledge—as opposed to electoral or authoritarian legitimacy—to exert and legitimate their rule, then the influence of technical expertise among decision-making circles is far from a new phenomenon in Chile.[54] The national-developmentalist model implemented after World War II, centered on a strategy of "import substitution industrialization" (ISI), was a result of concerted efforts by politicians and economists at the newly created Economic Commission for Latin America and the Caribbean (CEPAL).[55] The sort of economics that ISI-oriented economists practiced was, however, quite different from the current paradigm of neoclassical economic theory. Indeed, the ECLAC school of "structuralist economics" was underpinned by forms of statistical knowledge oriented

---

52 Parenti, "Environment-Making in the Capitalocene"; Andreucci et al., "Value Grabbing."

53 Parenti, "Environment-Making in the Capitalocene," 171.

54 Centeno and Silva (eds.), *Politics of Expertise*.

55 Marcus Taylor, *From Pinochet to the Third Way: Neoliberalism and Social Transformation in Chile* (London: Pluto Press, 2006); Silva, *In the Name of Reason*; Ossandón, "Economistas en la élite."

toward specific trajectories of empirical observation, despite being "scientific" in their own right.[56]

Patricio Silva shows how technical expertise has been a fundamental component of policymaking in Chile since the 1920s, when the governments of Arturo Alessandri and Carlos Ibáñez embarked on projects of state modernization at all levels.[57] Agencies such as the Chilean Economic Development Agency (CORFO), the Land Reform Corporation (CORA), and the National Planning Agency (ODEPLAN) were created during the mid-twentieth century as institutional centers for developing technical knowledge that could be applied to state-building and economic regulation.[58] Even during the short-lived socialist government of Salvador Allende, which according to commentators like Eduardo Silva was "hyper-ideological" and opposed to "technocratization," expert knowledge played a fundamental role.[59] In fact, Eden Medina has illustrated how, with the aid of a transdisciplinary and international group of experts, the Allende government designed a complex cybernetic system (known as "Project Cybersyn") for regulating the national economy in real time during the country's transition to socialism.[60]

It is thus historically inaccurate to view monetarist experiments in Chile during the Pinochet dictatorship as the first stirrings of technocratic rule in the country. Understanding such state-building endeavors as preceding the emergence of neoliberalism is important because it reveals how certain organizational cultures were to some extent already characterized by an "antipartisan" bent that favored technical knowledge as a fundamental component of institutional design and decision-making. This is one explanation for the overwhelming leverage enjoyed by economic experts since the Pinochet regime when implementing aggressive multiscalar economic reforms aimed at advancing neoliberal policy agendas, even after the transition to democracy.

The emergence of neoliberalism, however, was a turning point in technical experts' role in Chilean politics. Augusto Pinochet's regime

---

56   Ossandón, "Economistas en la élite."
57   Silva, *In the Name of Reason.*
58   Ibid.
59   Silva, *Technocrats and Politics in Chile.*
60   Eden Medina, *Cybernetic Revolutionaries: Technology and Politics in Allende's Chile* (Cambridge, MA: MIT Press, 2014).

arrived at the precise moment when the postwar edifice of ISI and state developmentalism was starting to collapse; a strong anti-interventionist and anti-inflationary policy consensus among ruling elites was emerging in the global North. In terms of economic theory, this period also corresponded to a shift from empirical and statistical observation of specific case studies to more abstract, mathematized forms of economic theory based on models aimed at universal validity. This was the background that allowed the Chicago Boys to be appointed architects of the military regime's institutional framework in ways that went beyond the purely economic.[61] An ordoliberal view of the free economy as *a political practice* was the underlying logic of this new modality of technical expertise. After Friedman made passionate claims about the urgency of implementing severe austerity programs to mitigate and amend Chile's then-ruined economy at a 1975 conference in Santiago de Chile, Pinochet appointed Sergio de Castro—a leading figure of the Chicago Boys—minister of finance.[62] Many of these young economists were also appointed at the ODEPLAN and by 1975 were already top advisors to the military in economic matters.[63]

The succession of events that followed is well-known and, as argued above, has become a totemic narrative of sorts in the history of neoliberal globalization. My aim here is not to revisit a well-trodden path, but to show how these experts influenced decision-making and electoral and parliamentary politics. According to Ossandón,[64] after the Chicago Boys, the role of economists transcended issues of development and macroeconomic policy and spilled over into broader political debates. The notion of technocracy broadened and economists' roles superseded that of technical consultants to include the holding of political

61   Alejandro Foxley, *Latin American Experiments in Neoconservative Economics* (Berkeley: University of California Press, 1983); Arturo Fontaine, *Los economistas y el presidente Pinochet* (Santiago: Zig-Zag, 1988); Silva, *Technocrats and Politics in Chile*; Carlos Huneeus, "Technocrats and Politicians in an Authoritarian Regime: The 'ODEPLAN Boys' and the 'Gremialists' in Pinochet's Chile," *Journal of Latin American Studies* 32, no. 2 (2000): 461–501.

62   Silva, *Technocrats and Politics in Chile*; Taylor, *From Pinochet to the Third Way*; Jessica Budds, "Water, Power and the Production of Neoliberalism in Chile, 1973–2005," *Environment and Planning D: Society and Space* 31 (2013): 301–18.

63   Budds, "Water, Power and the Production of Neoliberalism in Chile."

64   Ossandón, "Economistas en la élite."

positions, such as ministries and even presidencies.[65] This leads Huneeus to argue that neoliberalism emerged in Chile not just as a technical prescription for integral solutions but as a large, close-knit group of professionals who worked effectively to promote the success of the government.[66] This of course confirms Bonefeld's thesis that the difference between ordoliberalism and neoliberalism is one of nuance, not distinction.[67] Indeed, it is striking that ordoliberal diction crops up in Milton Friedman's contention that the state will "police the system [of private property], it will establish the conditions favourable to *competition* and prevent monopoly, it will provide a stable monetary framework, and relieve acute poverty and distress."[68]

The idea of the free economy resting on the state's use of organized force—the key tenet of ordoliberalism—is not exclusively confined to neoliberal reforms implemented in Chile; it emerges as an overarching hallmark of Latin America. According to Centeno and Silva, the broader context of political transformation across Latin American states at the onset of neoliberalism was marked by complex intermingling between political and economic domains.[69] For these authors, Latin America's new technocratic democracies still have elected representatives who exert nominal control over decision-making, but framing policy alternatives is largely in the hands of technical experts.[70] Significantly, the pivotal regulatory systems of neoliberalism in Latin America were enacted amid the authoritarian turmoil of military dictatorships. Neoliberal economic thought and the state's monopoly on organized violence came together as unlikely bedfellows. During the Chilean dictatorship, Huneeus explains, economic and coercive rationales went hand in hand, often taking an implicit—rather than explicit—form.[71] Most political and intellectual elites "opted not to know what was going on."[72] As an important member of the military regime stated in an act of overdue self-criticism, "the elite chose to 'live in a bubble.'"[73]

65 Ibid.
66 Huneeus, "Technocrats and Politicians."
67 Bonefeld, *Strong State*.
68 Ibid., 14.
69 Centeno and Silva (eds.), *Politics of Expertise*.
70 Ibid.
71 Huneeus, "Technocrats and Politicians."
72 Ibid., 473.
73 Ibid.

Understanding primitive accumulation as the constitutive premise of the existent economic forces, Bonefeld claims, "destroys their deceptive appearance as forces of nature"; it also refutes the idea that the fetishized economic forms have some abstractly conceived "human basis."[74] In the case of Chile, the idea of the rational, utility-maximizing individual (enshrined as the overarching tenet of statecraft) has concealed the modalities of expropriation that have delivered nature—human and nonhuman—to capital accumulation. At stake behind the technical expertise and algorithmic knowledge mobilized by the new economic intelligentsia in Chile, then, was an actual issue of property rights. Indeed, many of the economists who proposed the shock therapies implemented by the Pinochet regime, trained at the University of Chicago and other elite US institutions, also played important roles in Chile's largest business conglomerates.[75] Specifically, Manuel Cruzat (CEO of Grupo Cruzat-Larraín), Álvaro Saieh (media and communications tycoon), and Agustín Edwards Eastman (head of the Edwards group), whose corporations commanded salient sectors of the domestic economy such as banks, newspapers, transport, and mining, among others, became important brokers in the epistemic networks between the University of Chicago and Universidad Católica.[76] Also, many of the leading voices in Pinochet's economic team were appointed to the boards of directors and to executive positions in the corporate sector after the transition to democracy. Specifically, Pablo Baraona, Hernán Büchi, Carlos Cáceres, and Sergio de Castro had direct links with major energy and mining conglomerates such as Sociedad Química y Minera (SQM) and the Luksic Group.[77]

Revealing and condemning the conflicts of interest that result from neoliberal privatizations is of course a matter of huge relevance. However, pointing fingers at specific technocrats who used their privileged positions within the state apparatus to become rentiers would amount to the sort of Saint-Simonian moral reading of primitive accumulation discussed in the previous section.[78] The reading of primitive accumulation proposed by this chapter is sensitive to the use of

---

74 Bonefeld, *Critical Theory*, 84.
75 Fischer, "Influence of Neoliberals."
76 Ibid., 14.
77 Ibid., 42.
78 Bonefeld, *Critical Theory*; Roberts, "What Was Primitive Accumulation?"

concentrated force by landlords and state institutions, but places such extraeconomic mediations within the overarching framework of material interdependence begotten by value relations. As Sassen points out, the predatory nature of contemporary formations of capital does not spring exclusively from the fact that they include powerful elites.[79] The logic of capital accumulation is eminently systemic, which means that it could likewise function without the mediations of specific powerful actors. In its status as the alienated subject of the process of social reproduction, capital becomes constitutive of reified forms of subjectivity that are not exclusive to capitalists and workers. Politically progressive state officials and technocrats can also inadvertently embrace the forms of alienated economic thought that are instrumental to the "treadmill dynamic" of value. As Silva demonstrates, the opposition to authoritarian rule in Chile also adopted an increasingly technocratic character.[80]

Left-wing intellectuals, social scientists, and politicians during the military dictatorship were involved in think tanks and research centers that undertook critical studies of government policies and formulated alternative programs to be implemented after the transition to democracy.[81] The Corporation of Economic Research for Latin America (CIEPLAN) was a research center associated with the Universidad Católica's economics department, created in 1970 as an alternative to the Chicago Boys, who were becoming predominant within the university.[82] CIEPLAN split from the university, reopened in 1976, and concentrated on monitoring the economic policies of the Chicago Boys. After a few years, it became a fully fledged think tank of the Chilean Christian Democratic Party under the leadership of Alejandro Foxley.[83] Think tanks like these have become a structuring element of the cultural circuits of capitalism: they circulate ideas and specific cases and examples for the rest of the world—for example, Chile's pension system or Peru's regulation of property rights.[84]

CIEPLAN's economists, most of whom had doctorates in economics from top US schools and were immersed in transnational networks of

79   Sassen, "Predatory Formations."
80   Silva, *Technocrats and Politics in Chile.*
81   Ibid.
82   Ibid.
83   Silva, *Technocrats and Politics in Chile*; Silva, *In the Name of Reason.*
84   Ossandón, "Economistas en la élite."

expertise—became advisors to the government of Patricio Aylwin, the first civilian government after the Pinochet dictatorship. Aylwin's government, according to Silva, had a marked technocratic orientation, despite claiming to be politically progressive and left wing.[85] Indeed, Silva illustrates how several of the economic postulates of the Chicago Boys continued to influence policymakers in subsequent governments. The key economic principles introduced by the Chicago Boys were the need to relegate the state to a subsidiary role in economic matters, the primacy of foreign investment and of the private sector as drivers of development, the use of market mechanisms and efficiency criteria to allocate economic activities, and the need to keep public finances healthy. These principles remained untouched by the governments of the Concertación, the coalition that ruled the country for twenty years after the Pinochet regime.[86]

The self-proclaimed left accepted neoliberal economic principles because, according to Silva, first of all, the Allende government had failed at implementing a socialist economic model.[87] Second, despite some of its negative effects, many within the left thought neoliberalism had made the economy more efficient during the last three decades. A crucial missing element in explanations of this sort, however, concerns the broader role of the liberal state within the dynamics of capital accumulation. According to Roberts, part 8 of volume 1 of *Capital* proposes a model of the state that amalgamates the instrumental and parasitic.[88] The state is parasitic upon the accumulation of capital, which explains its formal independence from the actually existing class of capitalists as well as its imperfect instrumental relation to capital as such. The state under capital, Roberts argues, "is self-activating but subservient, a servile and corrupt henchman rather than an autonomous existence."[89] Roughly put, the state becomes addicted to capital, so technocrats cannot seek or even imagine an alternative path to that of endless growth and private ownership of the means of production. The "machine rationality" of neoclassical economics[90] and its tendency to consider transhistorical and unchangeable that which is historically determined make mainstream economic thought the perfect

85   Silva, *Technocrats and Politics in Chile*; Silva, *In the Name of Reason*.
86   Silva, *In the Name of Reason*.
87   Silva, *Technocrats and Politics in Chile*; Silva, *In the Name of Reason*.
88   Roberts, *Marx's Inferno*.
89   Ibid., 214.
90   Mirowski, *Machine Dreams*.

ally in producing the infrastructure and property systems that the mining supply chain requires. Understanding these rubrics of sociospatial design and engineering will be the aim of the next section.

Economists, Rentiers, and the Governance of Chile's Natural Resources

After Pinochet seized power and appointed several of the Chicago Boys to key decision-making positions within the Ministry of Finance and the ODEPLAN, they implemented an ambitious agenda of radical policy restructurings to reconfigure the institutional architecture of the Chilean state in almost every respect. These reforms, usually referred to as the "seven modernizations," involved the following: introducing new labor policies and legislation; transforming the social security system (especially pensions); municipalizing education; privatizing the health system; internationalizing agriculture; transforming the judiciary, and decentralizing government administration.[91] Despite the different aspects they addressed, these reforms were philosophically underpinned by the legacy of the marginalist revolution in economics and its subsequent revisions by the Austrian and Chicago Schools. In this worldview, economics has a rigorous scientific basis and provides a theory of price determination and resource allocation that is directly concomitant to the actions of a rational, utility-maximizing individual.[92] Such "methodological individualism" implies viewing private exchanges and market transactions as universal and therefore timeless. To the extent that each is aimed at improving the conditions of the parties in agreement, in aggregate terms, the "rational" and self-regulating market is viewed as a mechanism that serves the common good.[93]

Although much has been said about the reforms, programs, and "shock therapies" implemented by the Chicago Boys following free-market ideologies, only in recent years there has been an interest toward the socioecological and territorial implications of the neoliberal/

---

91   Silva, *Technocrats and Politics in Chile*; Taylor, *From Pinochet to the Third Way*.
92   Taylor, *From Pinochet to the Third Way*; Clarke, *Marx, Marginalism and Modern Sociology*.
93   Taylor, *From Pinochet to the Third Way*, 34.

neoclassical policy toolkit. This is surprising, because as Parenti points out, "managing, mediating, producing, and delivering nonhuman nature to accumulation is a core function of the modern, territorially defined, capitalist state."[94] For example, it has been argued that the Water Code enacted in 1981 was fundamental not only for the reconfiguration of the entire hydrosocial cycle, but also for consolidating the wider neoliberal program and the ambitions of its core supporters—i.e., business conglomerates, the military regime, and technocrats.[95] In addition, important reforms regarding energy production, landed property, agriculture, and mining allowed Chile to insert itself in the global economy and provided the robust legal and normative framework required for the sort of infrastructural systems that would later be developed as demand from Asian economies soared in the mid-1990s.

Mining and agriculture—the two flagship industries in Chile's primary-commodity sector—require access to large quantities of water, so the reforms to water rights were wide-ranging. The 1981 Water Code introduced a system of freely tradable private water rights and was underpinned by a logic of rational allocation whereby the market would redistribute scarce water to high value uses, because users would supposedly be incentivized to sell water if they did not need it.[96] This reform emerged from a process of "technical" revision of the existing water laws and regulations in 1979, which led economic experts to conclude that water rights needed to be converted into private property in order to transfer it from the state to users. This implied separating water from land (and from the embodied practices of social reproduction associated with it), and converting the former into a fully fledged commodity that could be freely traded in all sorts of markets—spot, financial, stock exchanges, and so forth.[97] Despite widespread opposition to these reforms, as Budds notes,[98] technocrats justified them on the basis of a convincing narrative framed around neoclassic economic principles

---

94   Parenti, "Environment-Making in the Capitalocene," 182.

95   Carl Bauer, "Bringing Water Markets Down to Earth: The Political Economy of Water Rights in Chile, 1976–95," *World Development* 25 (1997): 639–56; Jessica Budds, "Contested H_2O: Science, Policy and Politics in Water Resources Management in Chile," *Geoforum* 40 (2009), 418–30; Budds, "Water, Power and the Production of Neoliberalism."

96   Budds, "Water, Power and the Production of Neoliberalism."

97   Ibid., 306.

98   Ibid.

that illustrated the benefits of treating water rights independently from land, namely that secure property rights would create incentives for investment in water-related industries and water infrastructures and that water management would be transferred from the state to the users, creating more efficient allocation of resources, especially in arid regions.

During the second most important round of privatizations, between 1985 and 1989, nearly thirty state companies were transferred to private hands. Crucially, most of the privatized companies were from the natural resource industries and therefore highly reliant on water for their activities: transferring water rights constituted transferring valuable capital assets to capitalists.[99] Although the reforms that allowed these market developments to take place were portrayed as "value-free," based on technical knowledge, and oriented toward the common good, in the long run they consolidated the positions of the technocrats who formulated them after the transition to democracy, many of whom ended up holding executive positions in the privatized companies.[100] That is, some of the technocrats implementing neoliberal reforms became part of a new class of property owners and rentiers. Perhaps most problematic, Budds argues, after the transition to democracy, these economic principles regarding water provision and the regulations they underline remained completely unmodified.[101] There were attempts at introducing amendments, but they were always very controversial and reflected to a large extent the power relations embedded in Chile's elitist and polarized society. Within these debates, "technical experts"—who had strong connections to neoliberal think tanks and business conglomerates— delegitimized any attempts at reforming the water-provision system by framing them as "ideological" and juxtaposing them with the optimal effects of the supposedly neutral and value-free market.[102]

Under these regulations, the extractive industries can accumulate water rights and secure free and permanent water provision.[103] Also, application

99   Ibid.
100   Fischer, "Influence of Neoliberals"; Budds, "Contested H$_2$O"; Budds, "Water, Power and the Production of Neoliberalism."
101   Budds, "Water, Power and the Production of Neoliberalism," 310.
102   Ibid.
103   Manuel Prieto and Carl Bauer, "Hydroelectric Power Generation in Chile: An Institutional Critique of the Neutrality of Market Mechanisms," *Water International* 37, no. 2 (2012): 131–46.

processes for new water rights are complex and entail considerable legal knowledge, which often imposes insurmountable entry barriers to small-scale, disadvantaged market actors. Budds illustrates this through the example of ENDESA, a Spanish hydroelectric-power company that acquired almost all of the nonconsumptive water rights in southern Chile, preventing the entrance of competitors.[104] According to Prieto and Bauer, ENDESA owns 55 percent of all nonconsumptive water rights in all of Chile, and 98 percent of those granted within Region IX, the region of southern Chile with the highest hydroelectric potential.[105] In northern Chile, mining corporations acquire water rights for potential future mining projects and hoard them, denying smallholders access to water.[106] With this, Budds concludes, the water provision system has translated into a legal mechanism for expropriation among lower income groups and a catalyst for the untrammeled expansion of mining, export agriculture, and hydroelectric power generation through the accumulation and hoarding of water rights.[107] Although the force of lawmaking violence divorces labor from its means of subsistence, Bonefeld argues that it later appears in the "law" of value in the form of economic compulsion.[108] Dialectical thinking is therefore the instrument by which the violence of separation can be dissected from economic processes that present themselves as ahistorical, impersonal, and unchangeable.

The model of water provision designed by Chile's economic intelligentsia had far-flung implications. Simplistic representations and descriptions of this water scheme, says Bauer, have made technical experts and economists from the World Bank and the Inter-American Development Bank endorse its principles, presenting it as a successful model for international water reforms.[109] In other words, it has become

104 Prieto and Bauer, "Hydroelectric Power Generation."

105 Prieto and Carl Bauer, "Hydroelectric Power Generation," 136–37.

106 Jessica Budds, "Power, Nature and Neoliberalism: The Political Ecology of Water in Chile," *Singapore Journal of Tropical Geography* 25, no. 3 (2004): 322–42; Budds, "Contested H$_2$O"; Budds, "Water, Power and the Production of Neoliberalism."

107 Jessica Budds, "Water Rights, Mining and Indigenous Groups in Chile's Atacama," in Rutgerd Boelens, David Getches, and Armando Guevara-Gil (eds.), *Out of the Mainstream: Water Rights, Politics and Identity* (New York: Routledge, 2010); Budds, "Water, Power and the Production of Neoliberalism."

108 Bonefeld, *Critical Theory*.

109 Carl Bauer, *Siren Song: Chilean Water Law as a Model for International Reform* (Washington DC: Resources for the Future, 2004); Prieto and Bauer, "Hydroelectric Power Generation," 134.

a blueprint for market-driven allocation of water resources in Chile and beyond, which has profoundly favored the expansion of resource extraction across the continent. This casts the sociomaterial afterlives of economic models in stark light, especially considering how methods and instruments for rational calculation are increasingly imbricated in landscapes and geographies. In the English parliamentary enclosures, the methods of legal regulation, taxation, land surveying, and of course militarized evictions were necessary for capitalist industrialization to fully take off. According to Marx, such methods "conquered the field for capitalist agriculture, incorporated the soil into capital, and created for the urban industries the necessary supplies of free and rightless proletarians."[110] In a similar vein, the standards of measurement, rational calculation, and legal regulation that this section describes regarding electricity generation, water provision, and land tenure have also made possible the incorporation of Chile's vast natural wealth into the sprawling trans-Pacific system of mineral trade.

The regulatory framework for power generation enacted throughout neoliberal rule in Chile is an important example of the intrinsically political and ideological nature of a supposedly neutral policy apparatus. Through an analysis of the 1981 Water Code in conjunction with the General Law of Electricity Services (LGSE in Spanish) of 1982, Prieto and Bauer conclude that in Chile, fresh water is not allocated naturally by apolitical markets. Rather, it is channeled to extractive uses and privileged hydropower generation by an institutional framework that prioritizes industrial activities at the expense of other property rights. They show how almost none of the nonconsumptive water rights currently in use for hydropower generation were acquired through markets; they were acquired through privatizing state companies or the system of "original acquisition" implemented by the legal frameworks analyzed.[111] Furthermore, a reform to the 1981 Water Code implemented during Ricardo Lagos's presidency in 2005 created further incentives for resource extraction and hydropower production. Unused or underused water rights were taxed to incentivize further hydroelectric power production and reassert the condition of

---

110   Marx, *Capital*, Vol. 1, 895.
111   Prieto and Bauer, "Hydroelectric Power Generation."

water as a factor of production with specific relevance for the extractive industries.[112]

Prieto and Bauer note how, as a result of the prohibitive cost of large-scale storage of electricity, the LGSE created a sophisticated system to match power supply with contingent demand.[113] Under this framework, the first generator whose electricity is loaded into the grid is the one that can offer the lowest operating cost at that particular moment. This model was designed to avoid subjective political judgments that translated into preferences for specific power plants, companies, or technologies that could alter the equilibrium of an already complex and competitive market. However, such institutional arrangements are fundamentally biased: they embody preferences that make hydropower cheaper than other sources as they do not consider the downstream externalities it creates. According to a member of the Latin American Observatory of Environmental Conflicts (OLCA), since price is the only criterion that entitles providers to supply energy, market actors that specialize in clean power production (wind, solar, etc.) are automatically excluded because of the higher costs of their energy.[114] As a result, "dirty" forms of energy (from nonrenewable sources) have become increasingly predominant in Chile's energy matrix since the turn to neoliberalism. Besides hydroelectricity, further regulations have been implemented to intensify thermoelectric power generation as a means to cope with growing demand from the mining sector. This shift in the energy matrix has also been systematically propelled through layers of state policies designed for "efficient" allocation of scarce resources.

Mining activity has emerged as a crucial transformative force in Chile's material economy, provided that we understand the latter in terms of a geographically established process of resource and energy flows.[115] Because of its geological composition, the Andes mountain range is an immense source of mineral wealth that has shaped the identities and histories of South American societies over centuries. Chile's national territory contains possibly the largest portion of the Andes, and mining activity since the turn of the twentieth century has flourished.

---

112   Ibid.
113   Ibid.
114   Interview with Lucio Cuenca, director of OLCA, December 23, 2013.
115   Bridge, "Hole World."

With 13 percent of GDP, 60 percent of total exports, and 25 percent of fiscal income, mining is one of the foremost revenue sources in Chile.[116] Chile went from supplying 30.1 percent of the world's copper in 1995 to 47.5 percent in 2004.[117] During the 2010s, it has been implementing a set of strategies aimed at attracting further investment in mineral extraction, which means that mining has intensified substantially. Besides copper, the country has diversified its mining matrix and is beginning to supply gold, silver, titanium, and lithium, among other metals, to international markets.

Most importantly, the growth of the mining sector in Chile is directly related to reforms designed and implemented by economic experts during the Pinochet dictatorship. Through the National Law of Mining Concessions (LOCCM in Spanish) and the Mining Code, policymakers created a comprehensive set of incentives for transnational capital to develop mining projects, one of the most important being the mining concession.[118] The mining concession became a broadly used legal mechanism, especially after the 1990s, when most Latin American mining codes began to be reformed to liberalize domestic market conditions and attract investment for resource extraction. This was the case in Bolivia and Uruguay in 1991, Brazil in 1996, Venezuela in 1999, Mexico in 1992, Cuba and Argentina in 1995, Honduras in 1998, Nicaragua in 2000, and Colombia in 2001.[119] Although each local regulatory framework has its own particularities, Fuentes has identified three core characteristics that cut across all mining codes: First, the state is the sole owner of natural resources; property rights are neither imprescriptible nor subject to any statutory limitations whatsoever, although the general

---

116   COCHILCO, *Chile, país atractivo para las inversiones mineras* (Santiago: COCHILCO, 2013), 5.

117   Leire Urkidi, "Movimientos anti-mineros: el caso de Pascua Lama en Chile," *Revista Iberoamericana de Economía Ecológica* 8 (2008): 63–77.

118   César Padilla, "Minería y conflictos sociales en América Latina," in Toro et al., (eds.), *Minería, Territorio y Conflicto en Colombia*; Cancino, "La dudosa fortuna minera de Suramérica."

119   Adriana Fuentes, "Legislación minera en Colombia y derechos sobre las tierras y territorios," in Toro et al., (eds.), *Minería, Territorio y Conflicto en Colombia*; Toro, "Geopolítica energética"; Luis Álvaro Pardo, "Propuestas para recuperar la gobernanza del sector minero colombiano," in Luis Jorge Garay (ed.), *Minería en Colombia: fundamentos para superar el modelo extractivista* (Bogotá: Contraloría General de la República de Colombia, 2013).

tendency in the region is for states to delegate extraction activities to privately owned companies through concessions. Second, the underground is owned by the state, regardless of any property rights over the surface. Third, mining has been invariably declared a "public-interest activity," which means that the state is entitled to expropriate privately owned territories.[120]

As Parenti notes, the state is inseparable from the process of transforming socioecological property: it creates entitlements and rights that enable and uphold the process of capital accumulation.[121] By issuing property frameworks, titles, and rights, Parenti highlights, the state constitutes property as a social form, and this renders territory susceptible to private appropriation.[122] Concession holders exert absolute power over the territories handed over by the state, producing classic enclave economies that, according to Bridge, are at the same time deeply integrated into the global economy and fragmented from national space.[123] Since concessions only give their holders the right to the subsoil, surface rights must be acquired through market transactions, negotiation, or compulsory purchase.[124] This often results in displacement, enclosure, or outright expulsion, where corporations constitute "virtual republics" within territories of extraction. In Chile, the concession allows mining corporations to acquire vast territories for an unlimited period to develop extractive activities. Under the scheme of "full concession," implemented through Law 18.097 of 1981, once a given territory has been handed over to a mining company, the state has no legal right to demand its restitution unless it pays the concession holder for all minerals contained in the deposits.[125]

Cancino illustrates how under the LOCCM, enacted in 1983, Chile unilaterally waived its rights to receive revenues in the form of mining royalties from transnational corporations extracting copper for more than twenty years.[126] It managed to attract significant flows of foreign

---

120   Fuentes, "Legislación minera en Colombia," 217.

121   Christian Parenti, "The Environment-Making State: Territory, Nature, and Value," *Antipode* 47, no. 4 (2015): 829–48.

122   Parenti, "Environment-Making State."

123   Bridge, "Hole World."

124   Anthony Bebbington, "Underground Political Ecologies," *Geoforum* 43 (2012): 1152–62.

125   Padilla, "Minería y conflictos sociales."

126   Cancino, "La dudosa fortuna minera de Suramérica."

direct investment aimed at developing mining projects, which resulted in massive urban and sociospatial transformations all across the northern part of the country, especially in cities such as Antofagasta, Arica, and Copiapó. The once abundant glacier runoff that used to supply Copiapó with fresh water has been depleted by intensive mining operations and the city is struggling to find solutions to a deepening environmental crisis. According to a recent study, the water imbalance in Copiapó is so dramatic that the recharge of its aquifer system is an estimated 4,000 liters per second, while granted water rights—the vast majority for mining activities—demand between 21,000 and 25,000 liters per second.[127] The mining royalty system that underpinned such large-scale resource depletion across the country ended in 2005 when, after an intense parliamentary debate, the payment of royalties was re-established as a legal obligation on behalf of concession holders.[128]

Chile's neoliberal policy toolkits for mineral extraction governance have become increasingly ambitious over the years, and to some extent have even surpassed in scope and breadth those designed by the Chicago Boys themselves. The binational mining agreement signed by Chile and Argentina—the first of its kind in the world—is a clear example of how supposedly "progressive" Concertación governments (now rebranded as Nueva Mayoría) have taken neoliberal sociospatial engineering to a higher stage. In 1997, the governments of Eduardo Frei in Chile and Carlos Menem in Argentina signed an agreement on "mining integration and complementation" to facilitate resource extraction across the two countries' Andean territory. The aim of this agreement, Razeto and coauthors suggest, is to do away with geopolitical limits and establish a new geoeconomic order that allows transnational mining and energy corporations low taxes, low royalties, fast-track bureaucratic procedures, and logistical benefits across 4,500 kilometers of border territory.[129] According to Salinas, far from being a standalone legal instrument, this agreement is the result of previous installments and efforts by both countries (especially Chile) to lower regulatory standards and create

---

127   noalamina.org/latinoamerica/chile/item/7451-la-batalla-del-agua-de-copiapo, accessed on January 14, 2015.

128   Cancino, "La dudosa fortuna minera de Suramérica."

129   Camilo Razeto, Daniela Soto, and Andrés Marconi, *Pascua Lama, IIRSA: acumulación por desposesión. El imperio contraataca* 2009.

further incentives to attract foreign direct investment for resource extraction.[130]

Salinas notes how, for example, only a few months after taking office, the newly elected Aylwin government passed a tax reform, Law 18.985 of 1990, that granted tax and other benefits to mining corporations.[131] It established an effective income tax by which companies were no longer required to pay an assessed income tax of 4 percent on their overall mineral sales but were only taxed on the basis of their *actual* revenues.[132] This has been particularly problematic for public finances because, in practice, corporations use sophisticated accounting and financial maneuvers to declare themselves as generating little, if any, revenue.[133] A further legal instrument of the binational mining agreement is Law 19.137 of 1991, which transferred untapped mineral deposits owned by Codelco (Chile's state-owned mining company) to transnational corporations.[134] According to Luna and coauthors, following the enactment of this law, more than 300,000 hectares of mining concessions were transferred to transnational capital over four years without the state receiving a single dollar in return.[135]

Furthermore, the binational mining agreement is circumscribed within broader transcontinental attempts to create spatially integrated systems that should intensify resource extraction and facilitate the movement of raw materials across borders. It has been argued that the territorial extent of the binational mining agreement largely corresponds with the Southern Andean hub of the Initiative for the Integration of South America's (IIRSA's) regional infrastructure.[136] According to an OLCA member, these agreements are mutually complemented by IIRSA's aspiration to create an integrated infrastructural framework to

---

130 Bárbara Salinas, *Implicancias territoriales del conflicto Pascua Lama*, unpublished undergraduate thesis submitted to Universidad de Chile, 2007, 48. Diego Luna, César Padilla and Julián Alcayaga, *El exilio del Cóndor: Hegemonía Transnacional en la Frontera* (Santiago: Observatorio Latinoamericano de Conflictos Ambientales, 2004).

131 Salinas, *Implicancias territoriales*, 49.

132 Luna et al., *El exilio del Cóndor*.

133 Salinas, *Implicancias territoriales*.

134 Ibid., 50.

135 Luna et al., *El exilio del Cóndor*, 12.

136 For an account of IIRSA, see Razeto et al., *Pascua Lama, IIRSA*; Kanai, "The Pervasiveness of Neoliberal Territorial Design."

move raw materials between the Atlantic and Pacific oceans and dramatically reduce transaction costs and turnover times.[137] One of the most cherished dreams of economists and entrepreneurs is overcoming the geographical limits imposed by the sheer magnitude of the Andes to create an interoceanic corridor for enhancing transnational networks of trade with Asian markets, according to members of civil society organizations.[138] Within the framework of IIRSA and the binational mining agreement, Chile has been developing local infrastructural programs specifically tailored to facilitate mobility across mining regions in Andean territory.[139]

Crucially, to the extent that landed property— in the form of concessions, leases, or acquisitions—is at the core of this sociospatial engineering, the issue of ground-rent is fundamental for understanding the sociospatial transformations that have followed such state mediations. Since the land is monopolizable and alienable, Harvey contends, it can be rented or sold as a commodity.[140] A land title, which can serve as a claim upon anticipated future revenues, according to Harvey, becomes a form of fictitious capital.[141] With this, Harvey notes how the land market begins to function as a branch—with some special characteristics, of course—of the circulation of interest-bearing capital.[142] This has been clearly evidenced at the Latin American level in the context of the

---

137   Interview with Lucio Cuenca, director of OLCA, December 23, 2013.

138   For an account of interoceanic fantasies in Latin America, see Japhy Wilson and Manuel Bayón, "Black Hole Capitalism: Utopian Dimensions of Planetary Urbanization," *CITY* 20, no. 3 (2016): 350–67. These sorts of utopian dreams and hubristic ambitions were also mentioned and analyzed in interviews with members of Semillas de Agua, OLCA, and Consejo de Defensa del Valle del Huasco, in November/ December 2013.

139   Interview with Lucio Cuenca, director of OLCA, December 23, 2013.

140   Harvey, *Limits to Capital*, 367. According to Charnock et al., Harvey's notion of monopoly rent is crucial for making sense of urban change under contemporary capitalism. For Harvey, it is the locational uniqueness of the site or the scarcity of buildings, natural resources, etc., that allows landlords and developers to engage in monopoly pricing and several forms of speculative investments. Greig Charnock, Thomas Purcell, and Ramón Ribera-Fumaz, "City of Rents: The Limits to the Barcelona Model of Urban Competitiveness," *International Journal of Urban and Regional Research* 38, no. 1 (2014): 198–217.

141   Charnock et al., "City of Rents"; Andy Merrifield, *The New Urban Question* (London: Pluto Press, 2014).

142   Harvey, *Limits to Capital*, 367.

commodity supercycle.[143] In Chile, the case of water rights clearly exemplifies this trend, as in some cases they have become mere financial instruments that are traded in several sorts of markets regardless of actual demand or use.[144] When the land begins to be treated as a purely financial asset, Harvey argues, land ownership assumes a properly capitalist form and underpins the expropriation that ensures the reproduction of capitalism.[145]

For Harvey, the land market determines the allocation of capital to land and thus "shapes the geographical structure of production, exchange, and consumption, the technical division of labor in space, the socioeconomic spaces of reproduction and so forth."[146] Previous chapters have reflected on how the configurations of landed property that emerge from these institutional frameworks have translated into dispossession, displacement, and proletarianization of local communities. At the national level, Taylor has illustrated how all the main extractive sectors such as mining, forestry, fisheries, and agroindustry exist in the form of oligopolistic markets where a few large firms dominate.[147] These market-dominating firms, he adds, exercise considerable price-setting capabilities and enjoy generous state benefits. Although Taylor does not mention land tenure as a structuring element of such oligopolistic markets, it is clear that concession schemes, water rights, and electricity-production frameworks strongly determine concentration in the ownership of the factors of production. In sum, the logic of expulsion is inscribed in the category of rent to such an extent that Derek Kerr proposes to understand it as "the necessary form through which capital appropriates and commands space while at the same time enforcing labour's exclusion from that space, thereby reproducing the commodity status of labour-power."[148]

---

143   Purcell, " 'Post-Neoliberalism' in the International Division of Labor."
144   Prieto and Bauer, "Hydroelectric Power Generation in Chile: An Institutional Critique of the Neutrality of Market Mechanisms."
145   Harvey, *Limits to Capital*.
146   Ibid., 373.
147   Taylor, *From Pinochet to the Third Way*, 126.
148   Derek Kerr, "The Theory of Rent: From Crossroads to the Magic Roundabout," *Capital and Class* 58 (1996): 59–88, 85.

Conclusion

When discussing how the relative surplus population is historically produced via fraud and force, Marx decries the double standard of liberal economists. The economic intelligentsia deem the "law" of supply and demand eternal and immutable. However, when adverse circumstances prevent the production of a surplus population and therefore the absolute dependence of the working class upon the capitalists, he contends that capital, "along with its platitudinous Sancho Panza, rebels against the 'sacred' law of supply and demand, and tries to make up for its inadequacies by forcible means."[149] This chapter has demonstrated that things are not particularly different in the present moment, when liberalism continues to outspokenly condemn violence and coercion but relies on violence and coercion performed by *others* to secure its conditions of existence. The case of the Chicago Boys—and of neoliberal technocracy—could not be more illustrative. I have demonstrated that it is not that the neoliberal technocracies of Chile embraced the violent dispossession of small producers, workers, and peasants carried out by the army and police. Rather, neoliberal economists were able to implement legal and regulatory reforms informed by principles of economic liberty only as a result of the genocidal violence the military regime unleashed against those perceived to be the enemies of "freedom."

Although scholarly and popular approaches to resource extraction usually tend to foreground questions of force and of dispossession as their central explanatory element, I have argued that such aspects cannot be properly understood without also seeking their inverted manifestation in the purportedly "neutral" and "value-free" categories of neoclassical economic theory. Mining is not historically specific, and for this reason it has evolved alongside the modalities of personal dependence most characteristic of premodern societies (chattel slavery, pillage, feudal extortion, conquest, and so forth). However, modern society is where indirect forms of dependence emerge as the fundamental driving force behind spaces of extraction. Paradoxically, the autonomized force of economic abstraction—much more than the warfare of feudal landlords, conquerors, and ancient kings—has made today's mines

---

149   Marx, *Capital*, Vol. 1, 793–94.

substantially more contingent on violence and expulsion than those of the past. To fully grasp the nature of this contradiction, then, this chapter has presented an alternative reading of primitive accumulation that disaggregates the agencies at work in the acts of expropriation and extraeconomic force that constitute capital. The neoliberal economists of Chile decided to "live in a bubble" while the military was busy creating the clean slate that the former required to implement new regulatory frameworks regarding mining, energy, water, and land tenure.[150] The kernel of the Marxian theory of primitive accumulation, according to Roberts, consists in showing that the capitalists' power does not grow from conquest and plunder; it originates from the fact that they are neither the conquerors nor the plundered.[151]

Marx's account of systematic primitive accumulation is fundamentally divergent from the moralistic socialists' critique of conquest and usurpation; they view capitalist society as a more advanced iteration of feudalism—a civilizational system cemented on force and interpersonal coercion. The implications of this reading are crucial for emancipatory politics. The violence of extraction does not have its main thrust in the individual acts of treacherous corporations, murderous landlords, or corrupt technocrats. Most of the economic theory that goes into the policy apparatuses that govern the territorial configurations of mineral extraction has been produced by individuals who genuinely believe in the efficiency and fairness of free markets. When describing the "predatory formations" of capital, Sassen highlights that even if we eliminated the mediations of owners, managers, and other power brokers, we would not ipso facto eliminate the predatory ramifications of scientific thought.[152] An important feature of such predatory formations, she claims, is that they are contingent on *systemic*, rather than elementary or individual, power grabs.[153]

Although Chile may be the "textbook case" that exemplifies the predatory formations at the heart of liberal social science and policy apparatuses, it nonetheless indicates a widespread tendency easily perceived in many other places. Through an ethnographic study of Colombia's

150   Huneeus, "Technocrats and Politicians," 473; Fischer, "Influence of Neoliberals."
151   Roberts, *Marx's Inferno*; Roberts, "What Was Primitive Accumulation?"
152   Sassen, "Predatory Formations."
153   Ibid.

geographies of land-grabbing, Ballvé, for example, concludes that economies of paramilitary violence are not anathema to projects of modern liberal statehood.[154] Liberal tropes of "institution building," "good governance," and the rule of law, usually tied to initiatives to make spaces governable and attract foreign direct investment, are remarkably compatible with—although eminently distinct from—ruthless expulsion and violence.[155] Finally, this chapter is important because it provides an account of the historical and institutional context of state mediations within the overall spatial dynamics of extraction in the twenty-first century. None of the large-scale territorial transformations described so far would have been possible without the foundational role of technical expertise and mainstream economic theory. In the next chapter, we turn to the analysis of finance and the material worlds it summons into existence. Like science and expertise, finance is one of the crucial vital forces that animate primary-commodity production.

---

154   Ballvé, "Everyday State Formation"; Teo Ballvé, "Grassroots Masquerades: Development, Paramilitaries and Land Laundering in Colombia," *Geoforum* 50 (2013): 62–75.
155   Ballvé, "Grassroots Masquerades."

6. MONEY
Debts of Extraction

Like the imaginary appetites of the social parasite, money is a purely aesthetic phenomenon, self-breeding, self-referential, autonomous of all material truth and able to conjure an infinite plurality of worlds into concrete existence.[1]

As with the stroke of an enchanter's wand . . . [debt] endows unproductive money with the power of creation and thus turns it into capital.[2]

Introduction

By suggesting that money is "the blood of the Commonwealth," Thomas Hobbes was one of the first modern philosophers to highlight the central metabolic role of monetary flows in the production and reproduction of social life—comparable only to the role that the circulation of blood plays in the life of the human body.[3] Money—either in the form of debt instruments or in its most liquid expression as cash—is on this basis also central to the metabolism of the spatially integrated infrastructural systems of the planetary mine. In fact, the very possibility for the mining industry to

---

1   Terry Eagleton, *The Ideology of the Aesthetic* (London: Blackwell, 1990), 201.
2   Marx, *Capital*, Vol. 1, 919.
3   Hobbes, cited in McNally, "Blood of the Commonwealth."

become increasingly capital-intensive, smart, horizontally integrated, and autonomous has been directly contingent upon the mediations of a complex network of financial actors, practices, and instruments. Generally speaking, finance capital provides the liquidity required for investment-intensive projects aimed at energy production, manufacturing, resource extraction, or logistical arrangements at multiple scales.[4] For Harvey, fictitious capitals (such as credit and paper moneys) must necessarily be created ahead of real accumulation.[5] Colossal investments in the built and unbuilt environments, such as those required for the territorial infrastructures of extraction, would be simply untenable without systematic engagements between physical producers and the financial system. Along this process, stock exchanges, mining firms, traders, institutional investors, and rural populations become interwoven in relations of mutual transformation with remote geographies that become rapidly industrialized and urbanized—and often also destroyed.

The *financialization* of capitalism, understood as the increasing predominance of financial instruments, practices, and mechanisms over the actual production of goods or services in order to yield profits, is considered a landmark in the history of the capitalist mode of production.[6] Although social scientists have become increasingly interested in exploring how states, natural resources, corporations, and households become financialized, French and coauthors argue that insufficient attention has been paid to the role of space, which is usually accorded a passive

4   Harvey, *Limits to Capital*; Merrifield, *New Urban Question*.
5   Harvey, *Limits to Capital*, 295.
6   John Bellamy Foster, "The Financialization of Capitalism," *Monthly Review* 58, no. 11 (2007): 1–12; John Bellamy Foster, "The Financialization of Accumulation," *Monthly Review* 62, no. 5 (2010): 1–17; Brett Christophers, "On Voodoo Economics: Theorising Relations of Property, Value and Contemporary Capitalism," *Transactions of the Institute of British Geographers* 35, no. 1 (2010): 94–108; Brett Christophers, "Anaemic Geographies of Financialization," *New Political Economy* 17, no. 3 (2012): 271–91; Greta Krippner, *Capitalizing on Crisis: The Political Origins of the Rise of Finance* (Cambridge, MA: Harvard University Press, 2011); Sarah Hall, "Geographies of Money and Finance I: Cultural Economy, Politics and Place," *Progress in Human Geography* 35, no. 2 (2010): 234–45; Sarah Hall, "Geographies of Money and Finance II: Financialization and Financial Subjects," *Progress in Human Geography* 36, no. 3 (2010): 403–11; Costas Lapavitsas, "The Financialization of Capitalism: Profiting without Producing," *CITY* 17, no. 6 (2013): 792–805; Costas Lapavitsas, *Profiting without Producing: How Finance Exploits Us All* (New York: Verso, 2013); Alex Loftus and Hug March, "Financialising Nature?" *Geoforum* 60 (2015): 172–75.

role.[7] Under late capitalism, where the real subsumption of territory to capital has reached a new pinnacle, studies of the sociospatial ramifications of financial and monetary systems need to be reinstated as a key priority. Every historical cycle of accumulation—possibly since the emergence of Florentine banking in the fourteenth century—has coevolved with its own modalities of high finance. Shedding light into the transnational monetary configurations that underlie the present phase of capitalist development, therefore, is a task of substantial relevance.

As fiat moneys became the norm after the abandonment of the Bretton Woods gold standard in 1973, debt emerged as the very lifeblood that animates the reorganization of the mining industry into global supply chains. In order to finance the megainfrastructural systems required to attract foreign direct investment for primary-commodity production (ports, highways, railways, airports, supply chain security, etc.), Latin American states have become increasingly indebted to international financial institutions and more recently to East Asian economies, which in recent years have emerged as the world's main creditor nations. Also, in order to cope with an escalating demand of raw materials and with declining grades of mineral deposits, mining companies have been compelled to recur to a wide array of debt instruments to garner the cash flow required to continually modernize their instruments of production. Finally, as a matter of sheer survival in a context of rapid proletarianization, dismantled subsistence farming, and urbanization, Latin American peasants who coexist with geographies of extraction have been pushed into mounting levels of personal and household debt.

None of these geographies of money can exist without the others. Although fragmentary and insular in appearance, they all coalesce to form an organic, interrelated whole. The purpose of this chapter is therefore to mobilize Henri Lefebvre's spatialized reading of the Marxian notion of totality in order to offer an account of the financialization of capitalism that explores the dense interconnections between the global financial system, the geographies of resource extraction, and the production of everyday urban environments in the Huasco Valley, a region in northern Chile that has been abruptly reconfigured into a mining

---

7   Shaun French, Andrew Leyshon, and Thomas Wainwright, "Financializing Space, Spacing Financialization," *Progress in Human Geography* 35, no. 6 (2011): 798–819.

district in recent decades. Given Lefebvre's lifelong insistence on practical activity and lived experience as the pivot of dialectical thinking,
Charnock suggests that his mode of critical thought parallels those of
authors in the Open Marxist tradition and should therefore be read
productively alongside them.[8] For Lefebvre, we must mobilize the
notion of totality to grasp the multiple relations of material coexistence
between the private, urban, and global levels of social life.[9] If there is no
insistence upon totality, Lefebvre warns, theory and practice accept the
real just as it appears: fragmentary, divided, and disconnected.[10] The
case of the Huasco Valley constitutes a clear example of the intricate
relations between levels, because with the arrival of Pascua Lama—one
of the largest gold-mining projects in the world—the international
financial system and local geographies of extraction became intermingled in relations of mutual transformation.

In particular, this case offers a grounded vantage point from which to
visualize how the self-generating powers of money, themselves the ultimate expression of capitalist impersonal domination, extend their
command over ecosystems and everyday urban environments. As
argued in a previous chapter, monetary relations and market-making
have historically constituted the medium by which capital breaks the
mold of rural and agrarian ways of life. Personal financial instruments
(credit cards, mortgages, payday loans) as well as corporate ones (derivatives, options, stock issuances) revolutionize peasants' modes of existence in manifold ways, extending the discipline of capital to its constitutive outside. For this reason, a further objective of this chapter is to build
on Neil Smith's invitation to understand financialization as a particular
modality in which the real subsumption of nature (including *human
nature*) to capital asserts itself.[11] Just as the real subsumption of labor

8   Greig Charnock, "Challenging New State Spatialities: The Open Marxism of
Henri Lefebvre," *Antipode* 42, no. 5 (2010): 1279–1303.

9   Henri Lefebvre, *The Urban Revolution*, Minneapolis: University of Minnesota
Press, 2003 [1970]; Lefebvre, *Critique of Everyday Life*, Vol. 2 (New York: Verso, 2008
[1961]); Kanishka Goonewardena, "The Urban Sensorium: Space, Ideology and the
Aestheticization of Politics," *Antipode* 37, no. 1 (2005): 46–71; Andrew Shmuely,
"Totality, Hegemony, Difference: Henri Lefebvre and Raymond Williams," in Kanishka
Goonewardena, Stefan Kipfer, Richard Milgrom, and Christian Schmid (eds.), *Space,
Difference, Everyday Life: Reading Henri Lefebvre* (New York: Routledge, 2008).

10   Lefebvre, *Critique of Everyday Life*, Vol. 2, 181.

11   Smith, "Nature as Accumulation Strategy."

strips the laborer of their individuality, Smith argued that the real subsumption of nature (through its capitalization and financialization) likewise strips the biogeophysical environment of its specificity. The sociomaterial attributes of mountains, rivers, and forests, as well as those of human intentionality, are emptied of their content as finance renders them universally equivalent via the price mechanism.

The first section briefly engages with Lefebvre's ideas of levels and totality with the purpose of providing analytical foundations to make sense of the interrelatedness between the private, the urban, and the global in the production of the geographies of financial capitalism. The second section explores financialization as a mediated expression of the real subsumption of nature to capital. This discussion is central for clarifying the monetization of social life that subsumes the human body as well as the nonhuman environment to impersonal and quasi-objective forms of economic dependence. A third section grounds the theoretical discussion with a focus on the corporate and financial strategies behind Barrick Gold—one of the world's largest gold producers—and its truncated multibillion-dollar project, Pascua Lama. With those things in mind, the fourth section explores the production of urban space and the transformation of everyday life that followed the irruption of transnational finance capitals advanced for resource extraction in the Huasco Valley.

Levels, Totality, and the Urbanization of Finance

The notion of totality, which underpins much of Western Marxism's epistemology of society, regards the social whole as a structure or system constituted by parts that interrelate with the system.[12] For Paulo Freire, the notion of totality aims at overcoming the pitfalls of bourgeois epistemologies that purport partial, fragmented, and focalized views of

---

12 Fredric Jameson, *Marxism and Form* (Princeton, NJ: Princeton University Press, 1971); Georg Lukács, *History and Class Consciousness: Studies in Marxist Dialectics* (Cambridge, MA: MIT Press, 1971 [1923]); Paulo Freire, *Pedagogy of the Oppressed* (New York: Continuum, 2000 [1970]); Martin Jay, *Marxism and Totality: The Adventures of a Concept from Lukács to Habermas* (Berkeley: University of California Press, 1986); Bertell Ollman, *Dance of the Dialectic: Steps in Marx's Method* (Urbana: University of Illinois Press, 2003); Shmuely, "Totality, Hegemony, Difference."

reality, striving to comprehend total reality as an interrelated whole.[13] Although the notion of totality has been widely used, Kanishka Goonewardena notes that most approaches usually underscore it in some variation of a crude "base-superstructure" model.[14] Lefebvre's contribution for contemporary understandings of totality is therefore fundamental, because his reading proposes a more nuanced and flexible configuration of "levels"—beyond base and superstructure—in the context of late capitalism's historical totality.[15] Key to his reading of everyday life within the concept of totality is therefore an attempt to challenge the limitations of Althusserian and structural-Marxist modes of analysis that failed to uncover the negated human content of seemingly supratemporal economic categories and "structures."[16] By mobilizing totality, Lefebvre emphasizes the Hegelian lineage of the Marxian critique but also puts sensuous human practice at the center of dialectical thinking.

In *The Urban Revolution*, Lefebvre views the social totality as the result of the myriad interactions and flows between what he termed global (G), mixed (M), and private (P) levels of social practice.[17] With this, his intention is to clarify the constitutive relations between everyday life (P-level), the urban (M-level), and global neoliberalism (G-level),[18] a relation that can also be mobilized productively to make sense of the complex geographies of finance. In fact, French and coauthors argue that the spatially differentiated effects of financialization constitute one of the most important gaps in the burgeoning literature on the subject.[19] For them, accounts of financialization have prioritized the nation-state as the foremost container of economic activity, overlooking other relevant sociospatial units such as the city, the region, and the household, all of them central in the production

---

13   Freire, *Pedagogy of the Oppressed*.
14   Goonewardena, "Urban Sensorium"; Kanishka Goonewardena, "Marxism and Everyday Life: On Henri Lefebvre, Guy Debord and Some Others," in Goonewardena et al. (eds.), *Space, Difference, Everyday Life*; Goonewardena, "Planetary Urbanization and Totality."
15   Goonewardena, "Planetary Urbanization and Totality."
16   Charnock, "Challenging New State Spatialities."
17   Lefebvre, *Urban Revolution*.
18   Goonewardena, "Marxism and Everyday Life."
19   French et al., "Financializing Space."

and reproduction of the international financial system.[20] Although Lefebvre never addressed finance specifically, he was nonetheless aware of the multilayered and multidimensional ramifications of political-economic structures and for that reason deployed his analytic of totality as an "epistemological sensibility" with the potential to capture the radical relationality of social life under capitalism.[21]

Throughout *The Urban Revolution* and the three volumes of his *Critique of Everyday Life*, Lefebvre was equally critical of the all-encompassing analytical scope of the political economist and the methodological attachments to immediacy prevalent in ethnographic studies. The analytical challenge, he claimed, is drawing connections between such disparate levels of social reality.[22] Lefebvre's thoroughly relational understanding of the world is therefore circumscribed to what Bertell Ollman refers to as the philosophy of internal relations, a tradition of dialectical inquiry that considers the relation between parts and whole to be in continuous evolution and codetermination.[23] By foregrounding process, becoming, and mutual constitution, Ollman illustrates how the concept of totality allows viewing how the whole expresses itself through the part and how constant flux between parts reconfigures the whole.[24] The very notion of financialization—typically understood as an alteration of trading, investment, and consumption practices in crucial agents of capitalist accumulation such as nonfinancial firms, banks, workers, and households—implies an internally related conception of social practice that dialectically links global networks of institutions to the immediacy of the home.[25]

---

20   Ibid.

21   Shmuely, "Totality, Hegemony, Difference."

22   Henri Lefebvre, *Critique of Everyday Life*, Vol. 1 (New York: Verso, 2008 [1947]); Lefebvre, *Critique of Everyday Life*, Vol. 2.

23   Ollman, *Dance of the Dialectic*.

24   Ibid., 140–41.

25   Mazen Labban, "Oil in Parallax: Scarcity, Markets and the Financialization of Accumulation," *Geoforum* 41 (2010): 541–52; Hall, "Geographies of Money and Finance II"; Annina Kaltenbrunner, Susan Newman, and Juan Pablo Painceira, "Financialization of Natural Resources: A Marxist Approach," paper presented at the Ninth International Conference on Developments in Economic Theory and Policy, Bilbao, Spain, 2012; Lapavitsas, "Financialization of Capitalism."

In contradistinction to liberal political philosophies that construe the individual fact as isolated and self-contained, Fredric Jameson notes that dialectical thinking foregrounds the network of relations where the item may be embedded.[26] With the Lefebvrean analytic of levels as a methodological insight, subsequent sections will illustrate how household-level microfinancial practices are internally related to broader transformations in the global economy. This approach assumes particular relevance against the backdrop of dominant strands of thought that naturalize specific levels of social reality. Thus, on the one hand, a neo-Latourian stream of scholarship develops microsociological explanations of how human and nonhuman agents/actors "perform" financial markets and reifies the interactional attributes of finance in place-based contexts at the expense of drawing broader connections with political and economic transformations.[27] On the other hand, French et al., illustrate how the international political economy prioritizes the nation-state as the foremost container of financial activity.[28] The question of how transnational networks of production and circulation of social wealth rely on finance to further their command over the quotidian rhythms and essential fabrics of everyday life is usually absent from the Scylla and Charybdis of these interpretations. To provide further elements that breach the gap between these disparate levels of social existence, our task is now to discuss finance and the real subsumption of nature.

Finance and the Real Subsumption of Nature to Capital

The shift from formal to real subsumption, as noted earlier, has historically enabled capital to transcend biological, geological, and even

---

26   Jameson, *Marxism and Form*.

27   Hall, "Geographies of Money and Finance I"; Brett Christophers, "Wild Dragons in the City: Urban Political Economy, Affordable Housing Development and the Performative World-Making of Economic Models," *International Journal of Urban and Regional Research* 38, no. 1 (2014): 79–97; Neil Brenner, David Wachsmuth and David Madden, "Assemblage Urbanism and the Challenges of Critical Urban Theory," *CITY* 15, no. 2 (2011): 225–40.

28   French et al., "Financializing Space."

temporal limits to capital accumulation. As the constitutive outside of capital can no longer be cast in purely territorial terms (due to the completion of the world market), the intensification of the labor process via technological change enables the continuous production of this outside, even within the physical boundaries of the capitalist system. Boyd, Prudham, and Schurman were the first authors to apply Marx's distinction of formal/real subsumption to (external) nature systematically.[29] For these authors, the real subsumption of nature refers to systematic increases or intensification of biological productivity with the purpose of accelerating turnover times of capital. Nature, they argue, is not merely appropriated but rather (re)made to work harder, faster, and better. With the real subsumption of nature, capital circulates through nature as opposed to around it and biological systems are made to act as actual forces of production.[30] Despite its relevant contributions, this account presents two major shortcomings: it only considers the real subsumption of nature applicable to biologically based systems and industries, and it reproduces an externalist conception of nature that excludes the subordination of the *human* species (also part of nature) to capital.

The former of the above considerations leads Neil Smith to contend that the real subsumption of nature should also be understood as encompassing other, nonbiological ways of capitalizing socionatural systems.[31] Finance, in his view, constitutes a key example of how nature is reinvented to work for capital. With the emergence of new markets in ecological commodities, mitigation banking, and environmental derivatives, Smith notes how nature is bundled into "tradable bits of capital." Just as the real subsumption of labor strips the worker of her specificity, the real subsumption of nature through its capitalization in financial markets also abstracts nature from its particularity.[32] This process of abstraction is perhaps most ostensibly manifested in derivative-pricing models, which require that all concrete risks (political, climatological,

29   William Boyd, Scott Prudham, and Rachel Schurman, "Industrial Dynamics and the Problem of Nature," *Society and Natural Resources* 14, no. 7 (2001): 555–70.

30   Boyd, Prudham, and Schurman, "Industrial Dynamics and the Problem of Nature."

31   Smith, "Nature as Accumulation Strategy."

32   Ibid.

monetary, and so forth) be calculated by a single metric.[33] In other words, derivatives markets have to be able to translate concrete risks into quantities of *abstract risk*.[34]

It is important to stress that the notion of financialization is widely contested in the literature, assuming different meanings and characteristics depending on the context and the theoretical framework. Some authors no longer consider labor the central locus of corporate profit, arguing that financialization implies "profiting without producing."[35] Accumulation, so the argument goes, has emancipated itself from actual production, and the antagonistic relation between capital and labor has been revamped into an antagonistic relation between corporate managers and shareholders. This disconnection between interest-bearing capital and its monetary basis, as Marx rigorously illustrated in volume 3 of *Capital*, is only *apparent*.[36] Recent critiques of the notion of financialization have therefore illustrated that, notwithstanding that the locus of profit has shifted to some extent, the production of revenues is by no means severed from the "real economy."[37] Although the dominance of financial interests has an indisputable influence upon industrial capitals, Loftus and March contend, the relation with production remains fundamental.[38] As subsequent sections will reveal, the latter approach is clearly reflected in the complex, convoluted relation between the financial dynamics of the mining company and its extractive operations in Chile.

The commodity supercycle is premised on the deployment of an

33 David McNally, *Monsters of the Market: Zombies, Vampires and Global Capitalism* (Chicago: Haymarket Books, 2011), 163.

34 McNally, *Monsters of the Market*, 163.

35 Lapavitsas, "Financialization of Capitalism"; Jonathan Nitzan and Shimshon Bichler, *Capital as Power: A Study of Order and Creorder* (New York: Routledge, 2009); Joseph Vogl, *The Ascendancy of Finance* (London: Polity, 2017).

36 Marx, *Capital*, Vol. 3, 727.

37 Labban, "Oil in Parallax"; Mazen Labban, "Against Shareholder Value: Accumulation in the Oil Industry and the Biopolitics of Labour under Finance," *Antipode* 46, no. 2 (2014): 477–96; McNally, *Monsters of the Market*; Loftus and March, "Financialising Nature?"; Martín Arboleda,"On the Alienated Violence of Money: Finance Capital, Value, and the Making of Monstrous Territories," *New Geographies* 9 (2017): 95–101; Ana Carolina Campos de Melo and Ana Cláudia Duarte Cardoso, "O papel da grande mineração e sua interação com a dinâmica urbana em uma região de fronteira na Amazônia," *Nova Economía* 26 (2016): 1211–43.

38 Loftus and March, "Financialising Nature?"

interconnected network of physical and technological infrastructures across the South American continent, tailored to facilitate the swift circulation of raw materials. This complex logistical apparatus was, from the very outset, driven by diverse modalities of debt. To begin with, the consolidation of Asian economies as international lenders—especially as providers of balance-of-payments financing for deficit countries— has created a new source of liquidity for state expenditures on megainfrastructure projects in Latin America. Ever since the early phases of the new international division of labor, Japanese banks, state agencies, and sectoral associations have been devising and aggressively promoting new forms of international investment with resource-rich countries and local mining firms. With these financial strategies, Bunker and Ciccantell explain, they became increasingly adept at reducing their capital exposure on large infrastructure projects and inducing resource-rich economies to incur debt to finance huge mines as well as dedicated ports and railways networks.[39] By the 1980s, Japan had managed to weave a vast logistical apparatus for the procurement of iron, coal, aluminum, and copper, which encompassed Australia, Canada, Venezuela, Indonesia, Chile, Papua New Guinea, and Mexico, among others. Such transnational spatial technologies, according to Bunker and Ciccantell, could only be made possible thanks to the swelling debts of resource-exporting nations and mining companies.[40]

The tendency of Latin American countries to recur to international debt instruments issued by Asian economies became more pronounced during subsequent phases of the new international division of labor, reaching its pinnacle with the new monetary architecture that emerged after China's "Go Out" policy in 1999 and the internationalization of the Industrial and Commercial Bank of China in 2008.[41] With massive balance-of-payments surpluses from its enlarged manufacturing capacity, China's foreign exchange reserves have become the largest in the world ever—at a staggering $3.5 trillion by 2013.[42] During 2010 and 2011, Chinese state-owned financial institutions had allocated loans for nearly $357.9 billion, more than the total loans issued during the same

---

39   Bunker and Ciccantell, *Globalization*.
40   Ibid., 208.
41   Stanley, "El proceso de internacionalización."
42   Helleiner and Kirshner, "Politics of China's International Monetary Relations," 1.

time period by the World Bank and the IMF combined.[43] The majority of the loans allocated by China to foreign governments and firms, according to Schmalz,[44] are directly linked to infrastructure, primary-commodity production, and/or energy projects. Moreover, Schmalz notes that the main users of Chinese credit have been countries with limited or no access to international financial markets, and that also specialized in raw materials exports, such as Brazil ($21.8 billion), Venezuela ($65 billion), Argentina ($15.3 billion), and Ecuador ($15.2 billion). According to Gallagher, China has provided around $119 billion in loans and lines of credit to Latin American governments.[45]

Physical producers in the extractive industries—mining, oil, and energy firms—have also developed systematic engagements with the credit system, and reoriented their corporate behavior and strategies toward financial and speculative operations.[46] Expected future scarcity of raw materials as a result of heightened demand by productive capital—often artificially fabricated or dramatized by financial actors—contributes to the speculative frenzies that have driven price increases across most raw materials during the last two decades. Publicly traded oil companies, for example, tend to exaggerate the size of their reserves to inflate and distort the price of their shares in stock markets.[47] The financialization of nature is clearly observed in terms of the growing share of commodity derivatives of total derivatives traded. According to Kaltenbrunner et al., between 2004 and 2007 the gross value of commodity derivatives traded on over-the-counter markets went from $176 billion to $690 billion, an increase in the relative share from 2.8 percent to 6.2 percent.[48]

Besides speculative frenzies, the financialization of nature also encompasses a substantial transformation in corporate behavior among

43   Stanley, "El proceso de internacionalización"; Schmalz, "El ascenso de China."

44   Schmalz, "El ascenso de China."

45   Kevin Gallagher, *Latin America's China Boom and the Fate of the Washington Consensus* (Oxford: Oxford University Press, 2016), 7.

46   Kaltenbrunner et al., "Financialization of Natural Resources"; Labban, "Oil in Parallax"; Labban, "Against Shareholder Value"; Julie Ann De Los Reyes, "Mining Shareholder Value: Institutional Shareholders, Transnational Corporations, and the Geography of Gold Mining," *Geoforum* 84 (2017): 251–64.

47   Anna Tsing, *Friction: An Ethnography of Global Connection* (Princeton, NJ: Princeton University Press, 2005); Labban, "Oil in Parallax."

48   Kaltenbrunner et al., "Financialization of Natural Resources," 12.

physical producers. This aspect indicates the compulsions by which human activity becomes inverted and transformed into the attribute of an abstract process. Like workers, financial capitals extend the discipline of capital to the capitalist, who is compelled to behave in certain ways to avoid succumbing to interfirm competition. In strictly economic terms, McNally notes, it is capital that rules, not capitalists; the latter are mere bearers (i.e., forms) of capital's imperatives.[49] Such altered, quasiconscious behavior from nonfinancial firms is circumscribed within a general tendency that has been labeled "shareholder value," by which corporate managers, in seeking to compensate for poor performance in physical production, have instilled a universal competition for financial results.[50] Since the 1990s, corporate managers have been increasingly oriented toward achieving short-term financial value for shareholders, even at the expense of the integrity of the firm in the long term.[51] Labban notes how, in the United States, the payout ratio (the ratio of dividends to net corporate profit) for oil companies went from an average 41 percent between 1963 and 1979 to around 50 percent between 1980 and 1989, then soared to 84 percent in 2008.[52]

The system of shareholder value is driven by the proliferation of financial coupon ownership among middle class households and workers, whose premiums are pooled and managed by institutional investors seeking to increase the value of those assets.[53] It is in this very circularity between financial institutions and households where we can begin to grasp the myriad relations of codetermination between levels of social reality. As representatives of pooled assets and packaged debt, institutional investors tend to be unyielding in their relations with corporate managers of the companies in which they own stock. This, as Froud and coauthors contend, leads to huge pressure over management strategies, jeopardizing long-term integrity within the firm in favor of short-term returns. This, as we shall see, is patently

49   McNally, *Monsters of the Market*, 265.

50   Julie Froud, Colin Haslam, Sukhdev Johal and Karel Williams, "Shareholder Value and Financialization: Consultancy Promises, Management Moves," *Economy and Society* 29, no. 1 (2000): 80–110; French et al., "Financializing Space."

51   Julie Froud, Colin Haslam, Sukhdev Johal, and Karel Williams, "Accumulation under Conditions of Inequality," *Review of International Political Economy* 8 (2001): 66–95.

52   Labban, "Against Shareholder Value," 484.

53   French et al., "Financializing Space."

reflected in Barrick Gold's corporate strategies.[54] When investment strategies are so stringently disciplined by the logic of finance, even as corporate managers are aware of the perils involved, it becomes clear that the real subsumption of nature to capital encompasses more than extrahuman worlds. For Smith, what is missing in traditional accounts of the real subsumption of nature is a recognition of the productive powers of cooperation that characterize the human species in the deepest sense.[55] It was the power of technological and social organization, Smith argues, that made the shift from formal to real subsumption possible in the first place.[56] Finance, and the sophisticated technologies and epistemological frameworks underpinning it, is nothing but the seizure and autonomization of socialized labor as a means to increase profit.

The role of finance in the real subsumption of nature, however, is reflected in how financial imperatives discipline the "biology" (i.e., human productive subjectivity) of corporate managers. According to Labban, the imperative of shareholder value has been catastrophic for the integrity of firms as well as for the workers involved in primary-commodity production.[57] This, Labban contends, is because instead of slowing down material production—as mainstream readings of financialization suggest—managerial strategies have aimed at intensifying it, albeit substantially degrading how it takes place. Mergers and acquisitions, plant closures and capacity reduction, divestments, outsourcing and offshoring, streamlining, and ultimately, job destruction, Labban argues, are the bread and butter of the restructurings that companies continuously implement in their pursuit for shareholder value.[58] Moreover, as agrarian studies authors have pointed out, debt has historically figured as a key mechanism by which value relations are harnessed to shatter the mold of agrarian existence, transforming free peasantries into indebted proletarians.[59] Recent

---

54   Froud et al., "Shareholder Value and Financialization."
55   Smith, "Nature as Accumulation Strategy."
56   Ibid.
57   Labban, "Against Shareholder Value."
58   Ibid.
59   Araghi, "Great Global Enclosure"; Julien-Francois Gerber, "The Role of Rural Indebtedness in the Evolution of Capitalism," *Journal of Peasant Studies* 41, no. 5 (2014): 729–47.

waves of global restructuring, Araghi notes, have involved the institu-
tionalization of agro-export regimes, the deregulation of land markets,
and the drastic reduction of farm subsidies, thus forcing millions of
rural petty producers into debt to compete with transnational food
corporations.[60]

In the aftermath of the 2008 financial crisis, the rural populations
made redundant through previous rounds of agrarian restructuring—
now part of the surplus populations that inhabit shantytowns and
other precarious settlements—were systematically targeted by firms
mobilizing new varieties of debt instruments.[61] Microcredit emerged
as one of the most lucrative industries after the financial collapse,
especially when the G20 heralded it as a "core development strategy"
for overcoming poverty and the global recessionary environment. This
class-based project is draped in ideologies of "financial inclusion" to
extend the discipline of monetary relations to the surplus population,
creating a global poverty industry that draws profits from 2.5 billion
impoverished workers.[62] With these methods of disciplining the workforce and profiting from the surplus populations, finance has therefore
become a means to extend the discipline of capital to the very life of
living labor.

Barrick Gold and the Financialization of Capitalism

The historical turning point in the international price of gold is usually
attributed to the collapse of the Bretton Woods agreement in 1973,
when the US Nixon government decided to suspend the convertibility
of the US dollar to gold.[63] As a result, the US dollar became a fiat
currency. The price of gold, which for almost thirty years had held
constant at around $35 per ounce, skyrocketed due to increased

---

60   Araghi, "Great Global Enclosure."
61   Soederberg, *Debtfare States*; Adrienne Roberts, "Gender, Financial Deepening,
and the Production of Embodied Finance: Towards a Critical Feminist Analysis," *Global
Society* 29, no. 1 (2015): 107–27.
62   Soederberg, *Debtfare States*.
63   Aurelio Suárez, "El oro como commodity (producto básico), especulación
financiera y minería a cielo abierto," in Toro et al., (eds.), *Minería, Territorio y Conflicto*;
Kaltenbrunner et al., "Financialization of Natural Resources."

international demand and market volatility after the collapse of the Bretton Woods agreement, peaking at $1,837 per ounce in 2011. Current prices have made gold mining a very troublesome business, because previously remote and abandoned mining sites have once again become profitable and are reopening.[64]

Barrick Gold has become one of the most important companies in the gold-mining industry, with operations on five continents and mineral reserves of 104 million ounces of gold, 888 million ounces of silver, and 14 billion pounds of copper as of December 2013.[65] Registered in the Toronto Stock Exchange, Barrick extracted 7.42 million ounces of gold in 2012 and reached an operating cash flow of $4.2 billion in 2013, becoming the largest gold producer in the world.[66] Barrick Gold has thrived on risk since its very beginning, in 1986, when Peter Munk—its founder—bought a Nevada mine that turned out to be the most profitable gold discovery in the world, pushing the company into a leading position.[67] Barrick's strategy still reflects the ambition that gave it notoriety in the 1990s, and the company has grown steadily through an aggressive combination of production and acquisition, currently boasting twenty-six gold mines and further projects under development in Australia, North America, South America, and Africa.[68]

Gold production has not been the only driver of Barrick Gold, nor of the mining industry's profits in general, as companies have increasingly relied on combining actual physical production with intricate financial strategies that—most notably—include systematically issuing corporate debt and hedging. Thus, during the last decade, the largest gold-mining corporations (Barrick Gold, Goldcorp, Newcrest and Newmont) generated $47.5 billion in operating cash flows.[69] Yet these companies spent

---

64   *New Internationalist*, September 2014.

65   Barrick Gold, barrick.com/company/profile/default.aspx (accessed 12 September 2014).

66   Barrick Gold, *Annual Report 2013*, barrick.q4cdn.com/788666289/files/doc_financials/annual/2013/Barrick-Annual-Report-2013.pdf; Vladimir Basov, "Top 10 Gold Companies in 2012," *Mining*, February 19, 2013, mining.com.

67   Tsing, *Friction*.

68   Dave Brown, "Top 10 Gold Producers," *Gold Investing News*, November 15, 2010, goldinvestingnews.com.

69   Daniel Oliver, "Gold Mining Stocks Are an Increasingly Attractive Opportunity," *Forbes*, July 2, 2012, forbes.com,; Julie Ann De Los Reyes, "Pascua Lama in Barrick Gold's Strategy," *Verdeseo*, December 24, 2014, enverdeseo.wordpress.com.

$68.5 billion—$43.4 billion on net capital expenditures. To cover this gap in their balance sheets, these companies opted to issue equity shares before stock exchanges, which increased by 117 percent.[70] To fund over-runs in operational costs, Barrick Gold in particular has repeatedly relied on a combination of internal financing, corporate debt, and share issu-ances, allowing it to maintain its leadership among physical producers. Such financially oriented strategies, De Los Reyes argues, led Barrick Gold to issue shares for $3 billion in 2009 and for $4 billion in 2013, the two largest equity offerings in the history of the Toronto Stock Exchange.[71]

The pervasiveness of financial logics in the company's corporate governance can also be observed in its increasing orientation toward shareholder value as a result of the pressures exerted by institutional investors. Due to the underperformance of stocks, these investors have pressured the management to abandon a growth-oriented strategy; in their view, it undermines the company's ability to deliver shareholder returns.[72] Therefore, the company has recently adopted a more conserv-ative approach to capital allocation, which it has chosen to encapsulate under the motto "returns will drive production, production will not drive returns."[73] With this, Barrick Gold begins to reflect the general trend among physical producers described by Labban whereby internal power structures tend to shift in order to align the interests of managers with those of shareholders, who prioritize short-term profits over long-term growth.[74] In the company's 2013 annual report, Peter Munk announced that a comprehensive overhaul of the business strategy was being implemented in order to shift the focus from "production growth to maximizing free cash flow and risk-adjustment returns."[75]

The role of the state as a mediating agent in the expansion of physical production has also been crucial. The case of Chile, for example, illus-trates how political strategies and economic policies implemented at the level of the state create the material conditions for investment frenzies that can lead to vast inflows of foreign capital. According to Taylor, one of the key reforms the Chicago Boys implemented during the military

70   De Los Reyes, "Pascua Lama in Barrick Gold's Strategy."
71   De Los Reyes, "Mining Shareholder Value."
72   De Los Reyes, "Pascua Lama in Barrick Gold's Strategy."
73   Barrick Gold, *Annual Report 2013*, 1.
74   Labban, "Oil in Parallax."
75   Barrick Gold, *Annual Report 2013*, 3.

regime was reconfiguring the relation between domestic accumulation and global capitals.[76] By contrast to national-developmentalist models that protected domestic productive capitals, neoliberal reforms prioritize capital in its money form rather than as production.[77] This shift from productive capital to money form was forged through a process of deregulating both trade and finance. Postauthoritarian governments continued to implement policies of financial deregulation and subsidies, creating a very favorable business climate for large mining corporations like Barrick Gold, which after the 1990s started to develop low-cost, largely profitable investment projects across Chile's national territory. In the subsection that follows, we shift the focus of attention toward the geographies of extraction through Pascua Lama, Barrick Gold's flagship project and one of the largest untapped gold deposits in the world.

Pascua Lama and Barrick Gold's Corporate Strategies

Located in the Andes at around 4,600 meters above sea level, amid millenary glaciers and primordial rocks and near the source of the Huasco River, Pascua Lama is one of the largest and most controversial gold-mining projects in Latin America. Set to be developed between Chilean (70 percent) and Argentinean (30 percent) territory, Pascua Lama is the first binational mining project in the world. Its appeal lies in the fact that its estimated gold deposits amount to 16.9 million ounces of gold and 594 million ounces of silver that would be extracted over an area of 16.5 square kilometers, at an annual rate of 800,000 ounces per year.[78] It was initially branded "the largest low-cost mine in the world."[79] Barrick estimated construction costs at around $1.1 billion, but numerous setbacks and complications made that figure grow more than fivefold. According to the company's annual reports, consolidated capital expenditures on Pascua Lama for 2000 to 2013 amount to $6.6 billion, and this figure is set to keep moving upward as the company faces even more obstacles from several fronts.

---

76   Taylor, *From Pinochet to the Third Way*.
77   Ibid.
78   Urkidi, "Movimientos anti-mineros"; Barrick Gold, *Annual Report 2002*, 30.
79   De Los Reyes, "Pascua Lama in Barrick Gold's Strategy."

Such cost overruns evince the fact that the relation between the sphere of corporate finance as a driver of revenues and the sphere of production is anything but severed. It also demonstrates that there is no unilinear relation of causality between levels, as the relations of production taking place in Pascua Lama have been passive recipients of power structures, yet have themselves determined much of what happens at stock exchanges and courts in Toronto and New York. Lefebvre's reading of totality challenges mechanistic-determinist explanations of socioeconomic transformations and, in line with the philosophy of internal relations, shows how the dynamic movement of parts also exerts a constitutive effect on the unfolding of the whole. Pascua Lama was projected as an open-cast mine, and experts and communities warned from the beginning that the effects on glaciers and rivers would be devastating.[80] The license was granted in 2006 with around 400 caveats, after nearly sixteen years of preliminary explorations.[81] Construction did not begin until 2009, when Barrick Gold started to incur serious cost overruns: from spending $202 million in 2009 to $724 million in 2010 and then all the way up to $1.9 billion in 2013.[82]

Setbacks and obstacles in corporate operations are very common because, as a result of restructurings, cost optimization strategies, and layoffs, the mine performs under very poor standards.[83] While executives in Toronto make every effort to perform financial alchemy to increase dividends to shareholders, the company's engineers and managers at the extraction site strive to reduce operations costs by all means possible, for the sake of "shareholder value." According to Salinas, serial layoffs have also been at the core of Barrick Gold's strategies for maximizing profit and disciplining the workforce. Because of this, the unionization rate in Pascua Lama has reduced dramatically.[84] As a result, the company has faced repeated lawsuits, investigations, suspensions, and fines of up to $17 million, greatly undermining corporate operations. The company hid this from shareholders until a 2013 Greenpeace report

---

80   Raúl Molina and Nancy Yáñez, *La gran minería y los derechos indígenas en el norte de Chile* (Santiago: Observatorio de Derechos de los Pueblos Indígenas, 2008).

81   *Reuters Latin America*, "Suspensiones y multas a proyecto minero Pascua Lama de Barrick en Chile," April 10, 2013, lta.reuters.com.

82   Barrick Gold, Annual Reports, 2006–2013.

83   Interview with a Semillas de Agua member, November 27, 2013.

84   Salinas, *Implicancias territoriales*, 111.

alerted them to numerous obstacles in Chile.[85] In 2014, institutional investors filed a class-action lawsuit against Barrick Gold before the Ontario Superior Court of Justice for $6 billion, accusing it of concealing relevant information on extractive operations.[86]

Most importantly, what Pascua Lama demonstrates is that, as Moore contends, financial markets are indeed powerful ways of organizing nature.[87] If anything, the "driven by returns" corporate strategy exerts chaotic effects as nature becomes subsumed to capital, with rivers polluted, glaciers destroyed, crops diminished, workers laid off and impoverished, and communities in continuous anxiety and distress. The urbanization of the Huasco Valley has not only fractured landscapes and built environments: social, cultural, and financial practices have been reconfigured. The remainder of the chapter illustrates this process of real subsumption as the infrastructural systems of extraction expand to pervade the very fabric of everyday economic and financial practices.

Financialization of Everyday Life in the Huasco Valley

Vallenar is the capital of Huasco Province and the main town of the Huasco Valley, demographically and administratively.[88] Its location is strategic in geographical terms because it is next to the Pan-American Highway, a 28,500-kilometer road network that cuts across Chile from south to north and leads all the way to the United States and Canada. It is also equidistant between the mountain range—where vineyard agriculture and Pascua Lama and other mining projects are being developed—and the coast—where power plants are interspersed with tailings dams and olive plantations. It is therefore the point of convergence and hub of population flows, economic activity, logistical operations, and

85   Greenpeace, 2013. Lo que Barrick Gold oculta a sus accionistas: la contaminación de Barrick Gold traspasa la Cordillera de Los Andes y llegará hasta tu bolsillo. Report issued August 2013.

86   David Paddon, "Barrick Named in Class Action Suit over Troubled Pascua-Lama Project," Globe and Mail, May 22, 2014, theglobeandmail.com.

87   Jason W. Moore, "Toward a Singular Metabolism: Epistemic Rifts and Environment-Making in the Capitalist World-Ecology," New Geographies 6 (2014): 10–19.

88   According to the National Library of Congress (BCN 2013), the estimated population of Vallenar in 2012 was 46,207.

political institutions. A member of the Council for the Defense of the Huasco Valley noted that, before the mining boom, Vallenar had a deeply entrenched rural identity that gravitated toward self-sustaining local economies based on small landholding, subsistence farming, and trade.[89] Although local communities tend to interpret their current situation against the backdrop of a supposedly harmonious past, the Huasco Valley's landscapes had already been shaped and reshaped before the commodity boom by large-scale rounds of capitalist territorial restructuring. These emerged from postwar national-developmentalist projects, *hacienda* land-tenure schemes, and colonial rule.[90]

Previous rounds of territorial restructuring are not comparable in intensity or scale to the ones associated with the commodity supercycle. Since the turn of the century, when inward foreign direct investment started flowing haphazardly in pursuit of raw materials to supply Asian markets, Vallenar began to undergo dramatic transformations at a very fast pace, becoming the centerpiece of an urbanization process that has stretched across all of the Huasco Valley in less than ten years, transforming the built environment and ways of living. Electronic devices, satellite television, retail stores, consumer culture, and several other artifacts and practices became part and parcel of everyday life in this erstwhile agrarian town.[91] Microeconomic distortions caused by inward flows of investment, a local state official argues, led to a state-driven boom in infrastructural development as new roads and local public works like sports facilities, schools, and parks were built.[92]

At the mixed level, territorial transformation and urban sprawl in Vallenar were also reflected in the housing market, which grew at a dizzying pace with price increases of up to 300 percent. Construction companies from Santiago, such as P.I., Santa Beatriz, and Inca, began to develop housing projects with around 700 houses each, including the

---

89 Interview with a Council for the Defence of the Huasco Valley member, November 28, 2014.

90 The *hacienda* was the foremost unit of mercantile agricultural production in colonial Latin America. It was based on feudalistic social relations and, although it subsisted to some extent after colonial rule ended, it was gradually superseded by industrialized agriculture in the nineteenth and twentieth centuries.

91 Interview with a Council for the Defence of the Huasco Valley member, November 28, 2014.

92 Interview with an official from Vallenar's Planning Department (SecPlan), December 3, 2013.

first high-rise housing project in Vallenar, which amounts to 300 flats.[93] According to an activist, such price increases resulted in intraurban displacement, as many local tenants were no longer able to afford rents and had to move either to the town's periphery or to other cities like Copiapó and Calama.[94] Possibly foreseeing incoming flows of well-paid floating populations working temporarily at mining and energy projects, local investors built hotels, restaurants, bars, residence halls, and private accommodations, all quite successful until mining projects began to face legal problems with courts and regulatory agencies. As a result, the town went from a growth rate of 0.4 to 2.6 hectares per year.[95]

Monetary flows driven by mining investments also attracted retail chains, which began to open branches in Vallenar. Their arrival attests to how different levels of social reality interpenetrate urban space and shape how it is produced, materially and representationally. Although retail stores determine the internal dynamics of households to a large extent by altering consumption practices, the framework of levels allows us to visualize how such relations of mutual constitution are internally related to processes taking place at broader domains of social practice. Thus, the proliferation of retail chains in Chile is circumscribed within a global tendency that took place from the 1990s onward and consisted of the expansion of debt and retail banking across diverse segments of mass consumption as a result of the retreat of the state in the provision of public goods and the intensification of international trade. Ossandón shows that there are more than four retail cards per bank card in Chile, a striking figure when seen in perspective with other countries where the same rate fluctuates between 0.25 (Colombia), 0.9 (United States), and 1.5 (Brazil).[96]

The proliferation of credit offered by retail stores in Chile also reflects the tendency observed by Lapavitsas at the global level,[97] whereby nonfinancial enterprises become increasingly involved in financial processes on an independent basis, without recurring to banks. In Chile,

---

93   Ibid.

94   Interview with a Council for the Defence of the Huasco Valley member, November 28, 2014.

95   Interview with an official from Vallenar's Planning Department (SecPlan), December 3, 2013.

96   José Ossandón, "Sowing Consumers in the Garden of Mass Retailing in Chile," *Consumption Markets and Culture* 17, no. 5 (2014): 430.

97   Lapavitsas, "Financialization of Capitalism."

this shift in corporate behavior became distinctly rooted in corporate governance structures when retail chains began to realize that issuing credit cards allowed them to eschew banks' intermediation fees and value-added tax, and was a very good business in its own right.[98] As such, they began to develop increasingly differentiated financial products to be consumed by the Chilean middle classes.[99] As authors developing the notion of "financial ecology" have argued, the financial system makes great efforts to project its networks into different types of financial landscapes, especially those made up of less privileged individuals and households at the margins of society.[100] According to Marambio, retail chains in Chile fit into this very logic, as their aim has been to "include" segments of the population that do not have access to finance under regular credit-scoring standards.[101] Students, housewives, retired workers, and even the unemployed are the market for these retail financial products and, since they are considered "high-risk," are charged higher commissions, fees, and interests.[102]

Accordingly, the types of financial practices that have emerged in Vallenar with the expansion of new mineral extraction frontiers are the result of complex processes of economic transformation that crisscross the experiential and the planetary. The forms of predatory lending that have been projected across the geographies of Vallenar, and the towns of the Atacama Desert more generally, reflect such complex intermingling of levels because the local population is usually underremunerated, outsourced, or self- or temporarily employed. As a result, retail firms that have opened branches in Vallenar are those renowned in Chile for having a low-income customer base and for targeting marginalized, "financially illiterate" segments of the population. Their stores offer household appliances, iPads, clothing, medicines, holiday packages, and, most importantly, credit cards, now part and parcel of Vallenar's rapidly shifting urban landscape.

---

98   Ossandón, "Sowing Consumers."
99   Alejandro Marambio, *Bancarización, crédito y endeudamiento en los sectores medios chilenos: tácticas de acceso, diferenciación social y espejismo de la movilidad,* thesis submitted to the Universidad de Chile, 2011; Ossandón, "Sowing Consumers."
100   Elvin Wyly, Markus Moos, Holly Foxcroft, and Emmanuel Kabahizi, "Subprime Mortgage Segmentation in the American Urban System," *Tijdschrift voor economische en sociale geografie* 99, no. 1 (2008: 3–23; French et al., "Financializing Space."
101   Marambio, *Bancarización.*
102   Ibid.

The configuration and reconfiguration of these geographies of consumer credit show the eminently uneven nature of capitalist urbanization, because the financial actors that followed the arrival of transnational mining specialize in "high-risk" consumers. By contrast, top-tier retail stores like Falabella, an interviewee noted, have been completely uninterested in Vallenar despite the region's rapid economic growth.[103] Specific data on how many households have become subsumed by personal credit as a result of their interaction with retail stores is not available, because retail stores have not been declared proper financial institutions and are therefore not required to report financial activity to regulators. However, as figure 8 illustrates, the amount of private credit allocated by banks—which is rigorously reported to regulatory authorities—can illustrate the extent to which local households and firms have been relying on credit.

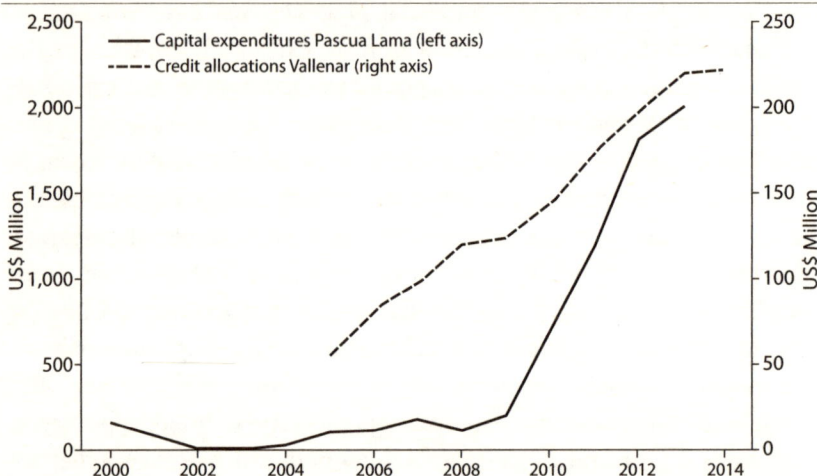

Figure 8 Credit allocations in Vallenar and Barrick Gold's capital expenditures on the Pascua Lama project

Source: Barrick Gold's annual reports and the SBIF's 2005–2014 annual reports on credit allocations. Barrick Gold's annual reports are available at the company's official website (barrick.com), and the SBIF's annual reports on credit allocations are available at the organization's website (sbif.cl).

Another striking trend in figure 8 is the correlation between capital expenditures by Barrick Gold in its Pascua Lama project and the amount of credit allocated locally by banks and other financial

103   Interview with an official from Vallenar's Planning Department (SecPlan), December 3, 2013.

institutions. With this, the interpenetration of levels that dialectically brings together corporate governance structures in Toronto and consumption and financial practices in geographically remote places like Vallenar start to appear not as ontologically distinct fragments but as parts of an internally related social totality. For Lefebvre, it is in everyday life where the interpenetrations between the natural, the everyday, and the historical are concretely realized.[104] Financial practices, especially in an institutional, corporate setting, are distinctively constitutive of the urban and in that sense, Vallenar has been urbanized by new infrastructures and built environments and by shifting frameworks of everyday activity. As such, financial intermediation and sales in hotels and restaurants are two further variables that also evince how finance and consumption increasingly predominate in shifting frameworks of interaction among the local community (see figure 9).

Figure 9 Financial intermediation in Vallenar and sales of hotels and restaurants in Vallenar

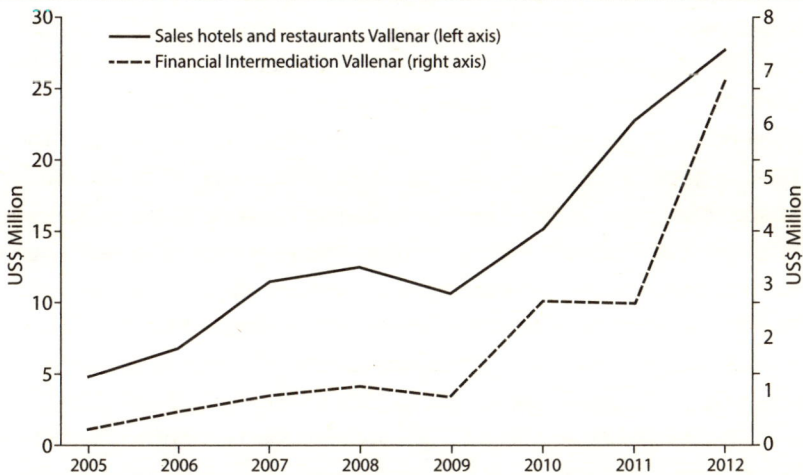

Source: Chile Internal Revenue Agency (SII), 2013 report of regional economic activity, sii.cl, accessed September 23, 2014.

One can easily see how local economies have been completely distorted by incoming flows of money simply by eating at a restaurant. During the heyday of the commodity supercycle, when metal prices

104   Lefebvre, *Critique of Everyday Life*, Vol. 1.

were at a record high, menu prices at restaurants and pubs in the town center were similar to those in central London. Financial intermediation is also rampant (see figure 9) and, given the sort of credit offered to the local community by second tier retail stores and other sorts of intermediaries, urban space is being produced in profoundly uneven and unjust ways. The networks of interaction established in space by these new segments of consumers and financial intermediaries are likewise constitutive of the emerging urban landscapes of Vallenar. Moreover, Schmid suggests that urbanization entails the enactment of networks of interaction and the diffusion and alignment of (urban) styles and fashions; credit cards, electronic devices, and fast food—among other cultural artifacts of urbanization—contribute to such shifting frameworks of interaction. As institutionally grounded financial practices and mass consumption patterns proliferate, agrarian identities and ways of living—which, before the turn of the twenty-first century, were deeply entrenched among the local community—start to erode and wither away as they become increasingly subordinated to the cash nexus.

Christian Schmid sees commuting as a clear example of how the urban landscape is produced through quotidian, situated routines.[105] The daily commute has been radically transformed by transnational mining and access to local credit, because the streets are now so full of cars that it is much easier to get around town on foot. Finding a parking space during working hours is extremely difficult. The total number of motor vehicles doubled during the 2010s.[106] Interviewees noted how, in recent years, traffic has become one of the most important problems, because Vallenar's streets are simply not suitable for so many motor vehicles. They also noted how the financial aspect of this outburst of urban traffic is deeply affecting the quality of life. As investment projects began to face legal complications and local labor markets became more precarious, the mindset of these newly indebted and urbanized subjects changed. The once enthusiastic car buyers started to feel besieged by interest from car loans as well as other forms of personal debt like mortgages, credit cards, and retail credit.[107]

105   Schmid, "Networks, Borders, Differences."

106   National Institute of Statistics (INE), Report on circulating motor vehicles during the 1984–2013 period.

107   Interviews with an official from Vallenar's Planning Department and two members of the Council for the Defence of the Huasco Valley, on December 3, November 27, and December 4, respectively.

Several interviewees frequently juxtaposed memories of their recent agrarian past with the current general anxiety caused by traffic-ridden streets, high rents, predatory lending, labor precariousness, and much of what is distinctive of everyday life in big cities. In an informal conversation, members of the local community remarked how they had been fooled into buying the "Santiago lifestyle": now they have to pay for everything, and everything they supposedly need is now sliced, diced, and sold at the nearest retail store. What they do not know is that it is not the Santiago lifestyle that they are living but the lifestyle of its most impoverished and marginalized neighborhoods, where financial provision is at its most predatory.

In a study of rural indebtedness, Gerber remarked how implementing credit/debit relations among the peasantry ultimately fosters market discipline, shaping capitalist rationality and culture.[108] By focusing on examples of rapidly industrializing countries, he noted how institutionalized credit was the transformative agent that eroded the "culturally and ecologically embedded "experiential" knowledge" of pre-capitalist societies, replacing it by an epistemological system that was characterized by an "algorithmic," rationalistic, and maximizing orientation.[109] What is particularly striking is that the shift in corporate behavior analyzed in previous sections, whereby short-term returns are increasingly favored over long-term growth, goes beyond large mining corporations and interpenetrates everyday practices of households and small-scale farmers. The real subsumption of nature to capital percolates to the most mundane aspects of everyday life. According to Gerber, once farmers enter an interest-based relationship, they are compelled to think and behave in particular ways to secure timely repayments.[110] The debtor begins to think in individual terms and prioritizes short-term benefits, "while surrounding sociocultural and ecological considerations remain secondary."[111]

Merely by chatting with members of the local community on an informal basis, it was palpable that credit relations have rendered the financialized, calculative, and profit-maximizing subjectivities Gerber

---

108   Gerber, "Role of Rural Indebtedness."
109   Ibid., 737.
110   Ibid.
111   Ibid., 738.

describes. Georg Simmel argued in an influential 1903 essay on the nature of cities that it was not precisely the built environment, population density, or infrastructural networks that defined the sphere of the urban. As the seat of the "money economy," Simmel wrote, the metropolis becomes a distinct sociospatial condition because it fills the daily life of many people with calculative mindsets that function in terms of profit-maximizing cost/benefit rationales.[112] Weighing, calculating, enumerating, and reducing qualitative value to quantitative terms, said Simmel, are practices constitutive of modern urban life.[113] The essentially individualistic character of the "mental life" of the metropolis is starkly juxtaposed with that of the small town, which rests more on feelings and emotional relationships.

A member of a civil society organization in Vallenar reported how, along with institutionalized credit systems, the proliferation of large-scale agribusiness has devastated agricultural production as well as smallholders' productive subjectivity. According to the interviewee, local producers feel increasingly pressured to implement profit-oriented monoculture techniques at the expense of long-term soil fertility. To increase productivity and compete with larger producers, they are also increasingly pressured to recur to debt instruments, widely available from all sorts of stores and financial intermediaries.[114] This calculative mindset, which for Simmel indicates the process of urbanization, has become widespread across the Huasco Valley, and banks; credit allocation in Vallenar attests to the degree to which finance has pervaded the most intimate confines of social life—family holidays, the daily commute, the purchase of medicines, etc. The extension of monetary relations into the most banal and mundane rhythms of everyday existence, it is worth stressing, signals mediations that transcend finance and become the embodied inscription of abstract, quantified time. For Postone, sociomaterial practices of quantification, cost/benefit analysis, and so forth, increasingly prevalent after the fourteenth century in Europe, were foundational as the commodity form asserted itself as the dominant structuring form of social life.[115]

---

112   Georg Simmel, "The Metropolis and Mental Life," in Kurt Wolff (ed.), *The Sociology of Georg Simmel* (New York: Free Press, 1950 [1903]).

113   Simmel, "Metropolis and Mental Life"; Mitchell, *Rule of Experts*.

114   Interview with a member of Creando Valle, December 1, 2013.

115   Postone, *Time, Labor, and Social Domination*, 213.

These reconfigurations in ways of organizing and exerting the process of social metabolism illustrate how, in general terms, monetary acceleration often translates into social acceleration, and how urbanization—in its modern, capitalist form—hinges upon the speed at which money circulates. If we think about the notion of totality in the context of these radical sociospatial transformations, then we begin to perceive how the indebted subjects of Vallenar reflect the complex dynamics and inner transformations unfolding within the mining industry, within resource extraction broadly considered and within global capitalism in general. The grim, geographically uneven financial landscapes of Vallenar are thus reminiscent of Lefebvre's claim that although it no longer makes any sense to think about city and country, it does not necessarily mean they have been harmoniously superseded, for they survive as places assigned to the territorial division of labor.[116]

Conclusion

This chapter has sought to demonstrate how contemporary spaces of extraction are crisscrossed by myriad monetary flows—cash and debt instruments such as derivatives, stock options, and even credit cards. The rapid urbanization of the Huasco Valley, one of Chile's main mining zones, illustrated how complex financial networks converge in the production of the fractured everyday environments of extraction. To begin with, the infrastructural grid that connects the Pacific Rim economies would have been unthinkable without the emerging modalities of sovereign debt geared at territorial upgrading via infrastructural development. Also, the financial strategies set into motion by corporate managers and other financial actors—via "shareholder value"—have been decisive for the mining industry to garner the liquidity it required to expand material operations at the extraction site. Barrick Gold and its flagship project Pascua Lama illustrate a trend that determines the practices and governance structures of the

116   Henri Lefebvre, "Space and the State," in Neil Brenner and Stuart Elden (eds.), *Henri Lefebvre: State, Space, World* (Minneapolis: University of Minnesota Press, 2009 [1978]).

mining industry as well as those of oil, forestry, and agribusiness corporations. As Barrick Gold's founder Peter Munk remarked in an annual report,[117] the "driven by returns" outlook has been implemented and followed by mining companies across the board and is now a blueprint for how to yield more returns to shareholders while simultaneously intensifying material production.

In avoiding fetishizing a single level of social reality as the dominant one, the chapter has also illustrated how the financialization of mining corporations supersedes mere material transformation of infrastructures and built environments and embeds itself in workers' and households' everyday frameworks of interaction. In revealing these expanding geographies of rural indebtedness and proletarianization, I have sought to show that the real subsumption of nature to capital is also concerned with the extension of the discipline of money to the daily rhythms of human existence. This case particularly illustrates how corporate managers, mining workers, and the local community have been disciplined in different ways by the logic of finance. At the level of corporate management, the unrelenting imperatives of shareholder value, mobilized by increasingly vocal institutional investors, thwarted executives' agency.

At the site of extraction, the possibility for worker mobilization has been dramatically hindered by constant pressure from layoffs, outsourcing, and streamlining. In the mining town, inward flows of capital expenditures from the mining company, added to the expansion of retail banking across landscapes of mass consumption, has translated into the financialization of culture and experience. The validation of value in the form of money, Bonefeld explains, is expressed in the apparent autonomization of the world market as the objective coercive force of modern society.[118] Indeed, the irruption of monetary relations across the quotidian landscapes of mining towns has subjected individuals and ecologies to impersonal forces that seem to lie beyond their control. This has become a source of social anxiety and frustration. However, while they beget an abstract mode of sociality, the monetary relations of primary-commodity production also actualize the material conditions for the collective power of labor

---

117   Barrick Gold, *Annual Report 2013*, 2.
118   Bonefeld, *Critical Theory*.

movements and subaltern groups. Understanding such budding syntheses of political and popular power and their articulation across the spatial distribution of the mining supply chain will be the task of the next chapter.

7. STRUGGLE

Plebeian Consciousness and the Universal Ayllu

It is necessary to study the connections that exist between the various parts which have been divided up by the sexual and international division of labor … But though these connections factually exist, they are almost totally obscured from our consciousness.[1]

Ch'ixi literally means marbled grey, formed by an infinity of black and white dots that become amalgamated in sense perception, but which remain pure, separate. It is a mode of thought, of parlance, and of experience underpinned by multiplicity and by contradiction. [2]

Introduction

In the Chinese context, the term *migrant* refers to the hundreds of millions of individuals who have left subsistence agriculture in recent decades to seek work in the booming industrial cities.[3] With the

---

1   Mies, *Patriarchy and Accumulation*, 3.

2   Silvia Rivera Cusicanqui, *Sociología de la imagen: miradas ch'ixi desde la historia andina* (Buenos Aires: Tinta Limón Editores, 2015), 295.

3   Mezzadra and Neilson, *Border as Method*; Eli Friedman, *Insurgency Trap: Labor Politics in Postsocialist China* (Ithaca, NY: Cornell University Press, 2014), 14; Qin Ling, "Introduction: The Survival and Collective Struggles of Workers in China's Coastal Private Enterprises since the 1990s," in Ren, Li, and Friedman (eds.), *China on Strike*.

establishment of the *hukou* system in the 1950s, the Chinese government created a tiered citizenship scheme for the provision of social services, where the population became conscripted to "rural" and "urban" localities. As a primary instrument of social distinction and control, rural migrants to the cities required permission to obtain employment, be allocated social services, or even marry or enlist in the army. The main consequence of this mechanism of demographic differentiation is that rural migrants have become second-class citizens, enabling the development of the hierarchical, cheap, and flexible national labor force that has propelled China's incorporation into the global economy since the 1980s.[4] To this day, migrants constitute the majority of the industrial workforce across China, fueling the construction and manufacturing industries. As of 2015, it was estimated that 247 million migrants were living in China's urban areas.[5] Moreover, Asia's geographies of large-scale industry generally rely more and more on women laborers, especially for highly standardized and menial tasks such as in most areas of electronics assembly, as well as parts and enclosures manufacturing.[6]

Composed mainly of former peasants and of racialized and gendered populations, the workforces that make up the East Asian industrial proletariat are strikingly similar in several respects to those of the mining supply chain in Latin America. As chapter 3 explained, the extractive frontiers associated with the commodity supercycle have expanded alongside the transformation of large peasant populations into cheap migrant labor. This trend, it should be pointed out, is not exclusive to the countries of the so-called global South. Recent accounts of the logistics supply chain highlight how the mechanization of port,

---

4   Cheng, "Hukou"; Shin, "Urbanization in China."

5   See Qiujie Shi and Tao Liu, "Glimpsing China's future urbanization from the geography of a floating population," *Environment and Planning A* 51(4) (2019), 817–819; see also Friedman, *Insurgency Trap*, 14.

6   McKay, *Satanic Mills or Silicon Islands?*; Lüthje et al., *From Silicon Valley to Shenzhen*, 154. In general terms, the feminization of labor has been a characteristic element of the new international division of labor ever since its earliest iterations. According to Maria Mies, in the 1980s, more than 70 percent of the labor force in the free trade zones of Southeast Asia, Africa, and Latin America, was female. The construction of an ideological vision of the non-Western woman as a bearer of cheap and docile labor-power has been actively promoted by "Third World" governments to attract foreign investment since the 1980s. Mies, *Patriarchy and Accumulation*.

shipping, and warehousing industries has developed in tandem with the racialization, outsourcing, and degradation of workers.[7] Transnational companies have actively harnessed considerations of citizenship, nationality, and ethnoracial attributes as a means to discipline labor and press down wages systemwide.[8]

The extended production of new racialized subjects, however, has precluded understanding of the growing material interdependence between individuals that directly underpins the technical configurations of production under late capitalism. In her landmark work on the early phases of the new international division of labor, Maria Mies points toward the increasing structural similarity between the production relations of the Western housewife and those of female agricultural and industrial workers in the so-called global South.[9] Although these groups of women had become materially connected via commodity production, each was oblivious to the existence of the other. This, Mies argued, precluded a transnational and anticapitalist political project that could challenge the historical thrust of patriarchy as well as its local, culturally specific manifestations. Likewise, while the emerging forms of sociopolitical contestation against resource extraction in contemporary Latin America are widely documented in the literature,[10] it is also important to explore such struggles' potential to connect with sociopolitical organization in other domains of social production. Superseding the methodological nationalism that informs most accounts of primary-commodity production, therefore, requires attention to how the lives of peasants and indigenous communities in geographies of extraction have become materially connected to those of workers in the manufacturing, logistics, and services industries.

This chapter thus draws connections between the struggles unfolding throughout the mining supply chain, mostly led by the peasant and indigenous communities whose lives are being radically transformed by debt, proletarianization, urbanization, and technological change. To

---

7 Tsing, "Supply Chains and the Human Condition"; Cowen, *The Deadly Life of Logistics*.

8 Ferguson and McNally, "Precarious Migrants"; Philips, "Migration as Development Strategy?"

9 Mies, *Patriarchy and Accumulation*.

10 Gago and Mezzadra, "Critique of the Extractive Operations of Capital"; Mezzadra and Neilson, "On the Multiple Frontiers of Extraction."

properly understand the transformative capacities of these emerging political actors, I build upon recent interventions that have unearthed the late Marx's partially published and little-known writings on precapitalist and non-Western societies. In these texts, Marx begins to question his erstwhile teleological worldview of revolutionary struggle and experiments with ideas about anti-imperialism and the multilinearity of history.[11] Overcoming his previous "unidirectional" view of international revolution, Marx became genuinely interested in tracing the relations between proletarian struggles in the metropolis and anticolonial movements in the colonies, Lucia Pradella argues.[12] Most importantly, these investigations allowed Marx to transcend the implicit superiority he had accorded to European societies as he began to identify in non-Western/premodern communities some of the latent yet unrealized elements of a future, more fully developed postcapitalist social form.

Yet the communal forms of social reproduction Marx observed in these unpublished investigations still could not be actualized as superior forms of social organization because their technical capacities confined them to an insular existence. In his view, it was the increase in the productive forces of society brought about by socialized forms of labor that would hence lay the premises for a new communal synthesis, one between the individual and "the human community" as a whole.[13] On this basis, the chapter advances a materialist approach to revolutionary subjectivity that avoids the pitfalls of cultural essentialism, and that grounds the basis of subaltern political agency in the immanent unfolding of capitalist social relations. For Postone, the conditions of possibility for critical, oppositional, or revolutionary consciousness are not to be rooted ontologically or transcendentally in the noncapitalist/precapitalist elements of social life. Without denying

---

11 Lucia Pradella, "Postcolonial Theory and the Making of a World Working Class," *Critical Sociology* 43, nos. 4–5 (2016): 573–86; Kevin Anderson, *Marx at the Margins: On Nationalism, Ethnicity, and Non-Western Societies* (Chicago: University of Chicago Press, 2016 [2010]); García Linera, *Forma valor y forma comunidad*; García Linera, *Plebeian Power*; Kojin Karatani, *The Structure of World History: From Modes of Production to Modes of Exchange* (Durham, NC: Duke University Press, 2014; Gareth Stedman Jones, *Karl Marx: Greatness and Illusion* (Cambridge: Harvard University Press, 2016).

12 Pradella, "Postcolonial Theory."

13 Karatani, *Structure of World History*; Anderson, *Marx at the Margins*, 159; Pradella, "Postcolonial Theory."

the existence or importance of such residual, noncapitalist tendencies, Postone claims that an approach that posits capital as the alienated subject of modern existence necessarily brings to light the transformative potentialities that are generated from within the framework of capitalism itself.[14] As subsequent sections illustrate, the encounter between subaltern politics and late-capitalist technology has triggered some of the most vibrant expressions of radical politics in both Latin America and Asia.

The chapter begins by engaging with Marx's ideas on the archaic community and his anthropological explorations of the multilinearity of history. I develop this discussion through an engagement with traditions of Latin American Marxism that have reflected critically on these insights, especially with respect to the complex relation between ethnicity and international socialism. The second section assesses the forms of political mobilization that have emerged in Latin America in the context of the commodity supercycle. In the third section, I reflect on the forms of labor insurgency that have unfolded in the onset of industrial upgrading and capitalist urbanization in Asia and reveal their similarity with those that gravitate toward resource extraction in Latin America. On the basis of this material framework of interdependency, the section then expounds a theory of revolutionary subjectivity that posits the necessity of transformative political action in the material transformation of the capitalist labor process. I argue that placing technological change at the center of contemporary theorizations of oppositional consciousness—rather than in a pristine, "uncorrupted," or abstractly free human essence—provides a means for grasping the unitary nature of struggles that are often viewed as separate, unrelated, and conditioned exclusively by culture. A final section grounds these discussions in an exploration of contemporary struggles against large-scale mining in the Huasco Valley. The chapter concludes by arguing for the need to overcome the fragmentation of human productive subjectivity imparted by the international division of labor, and to envision a trans-Pacific horizon of linked labor histories and geographies.

---

14   Postone, *Time, Labor, and Social Domination*, 38.

Decolonizing Class

Few aspects of the Marxian critique have been as controversial and misleading as the teleology of history that underpins a considerable part of its development as a critical-theoretical project. In initially espousing a mechanistic determinism that considered Western proletarian struggle to be of a higher order than that of subaltern and peasant communities in the fringes of the capitalist system, Marx has been rightfully portrayed as an unapologetic Eurocentric thinker. Moreover, his allegedly linear conception of world development, where the internal contradictions of feudalism give rise to bourgeois society, the latter of which set the foundations for a classless, communist society, has also garnered substantial criticism. According to Castro-Gómez, Marx inherited the Hegelian thesis of "peoples without history" and henceforth viewed non-European cultures as not yet capable of developing an economic structure that would allow their incorporation into a global revolutionary process with any measure of success.[15] Due to the semifeudal relations of many non-European societies in the nineteenth century, Marx viewed them as enclaves of counterrevolution on a global scale.[16]

It is well known that this interpretation became deeply engrained in decades of radical thought, especially as officially incorporated as a rationale of government in countries of the Eastern European bloc— "dialectical materialism," or the Soviet Diamat—during a considerable part of the twentieth century. The field of postcolonial studies therefore emerges as a strong reaction against such a deterministic and ethnocentric reading of history. Although postcolonial studies originated after the 1980s with the Subaltern Studies group of South Asian studies[17] and the Coloniality/Decoloniality group of Latin American studies,[18] it has gained prominence in geographical scholarship and urban studies more recently. Despite the internal differences between approaches and

---

15	Santiago Castro-Gómez, "(Post)Coloniality for Dummies: Latin American Perspectives on Modernity, Coloniality, and the Geopolitics of Knowledge," in Moraña et al. (eds.), *Coloniality at Large*.

16	Castro-Gómez, "(Post)Coloniality for Dummies," 262.

17	Ranajit Guha, *Elementary Aspects of Peasant Insurgency in Colonial India* (Durham, NC: Duke University Press, 2000); Dipesh Chakrabarty, *Provincializing Europe* (Princeton, NJ: Princeton University Press, 2007).

18	Quijano, "Coloniality of Power"; Mignolo, "Geopolitics of Knowledge."

methodological-normative assumptions, postcolonial studies in general is known for its critique of Eurocentrism, the geopolitics of knowledge, and economic reductionism. Its leading proponents see themselves as having excavated the sources of subaltern agency and reinserted culture as the central mechanism of social analysis.[19] Most importantly, their assessment is directed not only at the universalizing drive of Enlightenment thought but also at the Marxian critique, often construed as a radical offshoot of the colonial epistemological framework of the West. The Marxian notion of class, it is usually argued, obfuscates the modalities of social domination that underpin modern society and are understood as transcending the economistic focus of labor exploitation.

These increasingly popular fields of inquiry are therefore under-pinned by a strong assertion of the cultural specificity of the non-West and therefore assume that theoretical frameworks developed in the global North cannot grasp such cultural specificity. In tracing such hard and fast boundaries between West and East, global North and global South, the epistemological apparatus of postcolonial studies inadvert-ently reproduces the bifurcated world it sets out to criticize. Already in the 1990s, before so-called "identity politics" gained prominence in international academic circles, Ecuadorian philosopher Bolívar Echeverría—a renowned theoretician of cultural syncretism and capi-talist modernity in Latin America—was already puzzled by the self-defeating cultural essentialisms of emerging variants of Latin American indigenous politics. Pre-Hispanic cultures in the Americas, he argued, were undergirded by profoundly relational cosmologies and hence were unable to perceive the "Other" as independent alterity.[20] The otherness they perceived in the Spanish colonizers was for them a mere variant of the sameness of their own collective self.[21] The European mindset, on the other hand, always construed the cultural other in terms of radical alterity, as absolute negation of their own identity.[22]

---

19   Chibber, *Postcolonial Theory*.

20   Bolívar Echeverría, *La modernidad de lo barroco* (Mexico City: Era, 2010 [1998]), 29.

21   For Echeverría, their inability to construe the European invader as a cultural other gave indigenous peoples in the Americas a military disadvantage. This rendered them incapable of reaching a tendency toward absolute hatred and obstinate will to nullify the other.

22   Echeverría, *La modernidad de lo barroco*.

Echeverría was therefore perplexed to witness problematic expressions of inverted colonialism in the emerging identitarian movements of Latin America. Like the Cartesian mindset of European colonizers, the emerging political projects based on relentless defense of the virtues of the originary were also unable to seek, let alone even imagine, a different universalism.[23] In a similar vein, Silvia Rivera Cusicanqui challenges predominant readings of multiculturalism that dramatize ethnicity, rendering it exotic and thereby confining it within very narrow spatial and temporal limits.[24] These readings of indigeneity, she claims, reduce indigenous existence to inhabiting rural space and to a "theatrical" performance of cultural alterity almost invariably associated with the past. In this way, the indigenous political subject is stripped of its contemporaneity, thus ignoring its decolonizing impetus as well as its capacity to underpin an alter-globalizing project that can inform and also dialogue with the material bases of working-class politics in heavily populated urban areas.[25] The very rubric of "indigenous peoples" (a term that can be extended to the popular classes as well) she claims, "ultimately disavows the status of indigenous populations *qua* majorities, and therefore denies their hegemonic and state-making capacities."[26]

Recent studies of Marx's lesser-known engagements with non-Western and precapitalist societies have aimed in the direction of transcending an erstwhile "stageist" approach to history, but without replicating the inverted orientalisms and fragmenting visions that have precluded the theorization and practical construction of a subaltern futurism that is global, pluralistic, and revolutionary. During the later years of his life, Lucia Pradella notes, Marx overcame Eurocentric and unilinear views of development and recognized the seeds of material interdependence and collective power of an emerging world working class.[27] Although this shift of vision figures in a somewhat fragmentary manner in Marx's mature social theory, his systematic engagements with questions of ethnicity, colonialism, and non-Western societies are only partially published and translated. These

---

23 Ibid., 27.
24 Rivera Cusicanqui, *Sociología de la imagen*; Rivera Cusicanqui, *Un mundo ch'ixi es posible*.
25 Rivera Cusicanqui, cited in Verónica Gago, *La razón neoliberal: Economías barrocas y pragmática popular* (Madrid: Traficantes de Sueños, 2015), 86.
26 Gago, *La razón neoliberal*, 86.
27 Pradella, "Postcolonial Theory."

notebooks, some of them written between 1879 and 1882, are part of a vast thirty-two-volume corpus in the ongoing *Marx-Engels-Gesamtausgabe*.[28] Key in these notebooks are Marx's notes on Maxim Kovalevski as well as the letter he wrote to Vera Zasulich in 1881. Through analyzing agrarian societies in transition in Russia, Poland, India, ancient Rome, and to a lesser extent Latin America, Marx explored different paths toward social development beyond a mechanistic succession of stages. At this point, he began to insist on a site-specific, multilinear view of history that, far from being limited to the study of peripheral regions, is positioned to offer elements for an alternative path toward socialism.

Though he originally espoused Bernier's theory of "oriental despotism," where peripheral communities are deemed backward and irrelevant to revolutionary struggle, Marx gained much inspiration from Kovalevski's anthropological explorations on anticolonial struggle and communal landownership in India.[29] In distancing himself from the alleged inferiority of the non-West implicit in "oriental despotism," Pradella argues, Marx began to overcome the prevailing dualistic representation of East and West and laid the foundation for a unitary scheme of social development.[30] Most importantly, Marx began to see, in non-Western forms of landownership and social reproduction, some of the embryonic and unrealized forms of the universalized political community of the future. Furthermore, he began to identify the revolutionary potential of peasant and non-Western communities, especially in the context of anticolonial struggles at the fringes of European societies.

Although Latin American Marxism has been considerably overlooked in these new studies, I claim that it offers powerful tools to rethink the idea of the global working class beyond its Eurocentric bias. Due to the encroaching presence of "informality" and labor-intensive regimes of social production, several traditions of Marxist thought in Latin America have been driven by the necessity to conceptualize class relations beyond those of Western capitalism. The work of José Carlos Mariátegui, for example, was oriented by the aspiration to forge an "Indo-American" socialist political project that departed from the

---

28   MEGA II/IV; García Linera, *Forma valor*; García Linera, *Hacia el gran Ayllu universal: Pensar el mundo desde Los Andes* (Santiago: Editorial Arcis, 2014); García Linera, *Plebeian Power*; Anderson, *Marx at the Margins*.

29   Anderson, *Marx at the Margins*.

30   Pradella, "Postcolonial Theory."

evolutionism and mechanistic determinism prevalent in the main-stream Latin American left of his time, where indigenous and peasant struggle was considered backward and irrelevant.[31] The overarching thread in Mariátegui's work is a utopian-revolutionary dialectic that looks back to elements of the precapitalist past but systematically points forward to a socialist future.[32] Scorned by Peruvian nationalists for his "deviations" into European Marxism, as well as by Latin American corporatist socialism for his romantic populism, Mariátegui was considered an outcast in his time for trying to overcome the false dichotomy between universal and particular. As Webber notes, Mariátegui defended the universal character of the struggle for socialism, seeing it as a possibility opened up by global capitalist processes.[33]

The early writings of Bolivia's current vice president, Álvaro García Linera, also provide important insights for transcending the analytical and political pitfalls of cultural essentialism. Before his controversial role in the government of Evo Morales, García Linera had been the ideologue of a militant branch of Marxist indigenism (the Túpac Katari Guerrilla Army—EGTK). Frustrated with how the mainstream Bolivian left disregarded peasant and indigenous political mobilization, García Linera developed pioneering studies of the late Marx's interest in the ethnic question years before the topic began to arouse interest among Anglophone academia.[34] After joining the Morales government and orchestrating the state crackdowns on the indigenous political subject his earlier writings had praised, García Linera was ostracized by the Latin American intelligentsia. He is best known for his most recent intellectual work, where he adopts a posture that is radically at odds with his erstwhile multilinear explorations of world history.[35] In his recent and highly controversial book *Geopolítica de la Amazonía*,[36] the

---

31   José Carlos Mariátegui, *7 ensayos de interpretación de la realidad peruana* (Caracas: Fundación Biblioteca Ayacucho, 2007 [1928]).

32   Jeffery R. Webber, "The Indigenous Community as 'Living Organism': José Carlos Mariátegui, Romantic Marxism, and Extractive Capitalism in the Andes," *Theory and Society* 44, no. 6 (2015): 575–98; George Ciccarello-Maher, *Decolonizing Dialectics* (Durham, NC: Duke University Press, 2017), chapter 4.

33   Webber, "Indigenous Community."

34   Bruno Bosteels, *The Actuality of Communism* (New York: Verso, 2014).

35   Webber, "Indigenous Community."

36   Álvaro García Linera, *Geopolítica de la Amazonía: Poder hacendal patrimonial y acumulación capitalista* (La Paz: Vicepresidencia del Estado Plurinacional, 2012).

author adopts the teleological approach he fiercely criticized in the past, arguing that Bolivia will never establish true socialism without first building an industrial-capitalist base. On this basis, he has been unrelenting in his critique of indigenous and other environmentalist groups that reject the government's plans to expand natural resource frontiers, branding them as "infantile pawns" of foreign nongovernmental organizations or even the "useful idiots of imperialism."[37]

This is certainly unfortunate, because without in any way condoning his institutional practice as a state official, García Linera's early writings are some of the most lucid attempts at rethinking the Marxian critique beyond Eurocentrism. As a former student of Echeverría, García Linera reflexively explored the nuances and internal differentiations of the contemporary mode of production as it stretches beyond the heartlands of capitalism.[38] In Latin American economies, the process of real subsumption of labor to capital is far from being fully complete or even embedded within a linear temporality. He notes, for example, how the model of accumulation in Bolivia unites—in a tiered manner—the production structures of late modern capitalism, with "circuitous mechanisms of exaction and the colonial extortion of domestic, communal, artisan, *campesino* and small-business productive forces."[39] In the context of this "baroque modernity"[40] or "motley society,"[41] a rejuvenated technological-organizational basis of production and resource extraction has "consciously and strategically subordinated the informal shop, home-based work and the kinship-networks of the subaltern classes to numerically controlled systems of production ... and the monetary flows of foreign stock-markets."[42]

The sheer economic and cultural heterogeneity emerging from these sociotechnical systems, for García Linera, coexist as an amalgamated, baroque socioeconomic formation with definitive effects on what he refers to as the "multicivilizational" constitution of the political community. Reflecting on Marx's Kovalevski Manuscript as well as

37   Webber, "Indigenous Community."
38   García Linera, *Plebeian Power*.
39   Ibid., 212.
40   For an overview, see Echeverría, *La modernidad de lo barroco*.
41   René Zavaleta Mercado, *Lo nacional-popular en Bolivia*, Mexico City: Siglo 21 Ediciones, 1986.
42   Garcia Linera, *Plebeian Power*, 212.

on the letter to Zasulich, García Linera considered the Andean indigenous and peasant *ayllu* (i.e. a form of communal association in the Andean region, especially found among the Quechuas and Aymaras) a potential model and starting point for the universalized political community of the future.[43] Bringing these texts to life through an analysis of Latin American reality, he argued that an authentic insurgency against the domination of capital is simply unthinkable outside the communal and class struggle to universalize the communal social rationality that characterizes the *ayllu*.[44] Bosteels adamantly warns against interpreting this as the type of naïve pastoralism that romanticizes an idyllic past, or even as a developmentalist illusion of inevitable progress.[45] Instead, Bosteels argues that García Linera is actually pointing toward the fact that "it is only from within the contemporaneity of international exchange and the universalization of capitalism that simultaneously the possibility arises for a rearticulation of communism."[46] On this basis, what makes Marx's texts on the archaic community particularly generative is that they circumvent the culturalist trap of romanticizing a supposedly "pristine" essence of the subaltern subject and instead root the determinations of its political agency in the entanglements and interdependencies that underpin capital accumulation on a world scale. In the section that follows, we explore the implications of such an approach for making sense of the politics of large-scale mining.

The Universalism of Difference

According to Fredric Jameson,[47] one of the most salient features of the utopian imagination in the era of globalization is the fact that it has entailed a shift from the idea of a singular utopia toward *utopias* in many forms. These decentralized modalities of political thought and practice, he argues, stand opposed to earlier narratives of communication,

---

43   García Linera, *Forma valor*; García Linera, *Hacia el gran ayllu universal*.
44   García Linera, *Plebeian Power*, 51.
45   Bosteels, *Actuality of Communism*.
46   Ibid., 252–53.
47   Fredric Jameson, *Archaeologies of the Future: The Desire Called Utopia and Other Science Fictions* (New York: Verso, 2007).

hybridization, and multiculturalism. Such autonomous and noncommunicating worlds of social practice are therefore conceptualized by Jameson in terms of a *utopian archipelago*: a textured system of discontinuous centers that combines properties of isolation with those of relationship. Generally speaking, the idea of the utopian archipelago resonates with recent efforts to reveal the intrinsically generative dimensions of contradiction between opposites, and between difference generally considered.[48]

Autonomous domains of subaltern and anticapitalist social practice, however, encapsulate a paradox that undermines their political effectiveness, especially regarding their capacity to jump scales and articulate a broader emancipatory project. In giving consistency to a territorial reality at odds with capitalist modernity, autonomous units can become isolated and unable to expand their radical societal forms. This is what, in Marx's view, thwarted the possibility for the ancient community to become a universal-superior form of social organization. Ancient communal forms, García Linera argues, do not propose universalizing human bonds as the default mode of social reproduction.[49] Such contradiction in the ancient communal form, in his view, could be overcome as a result of the sociomaterial powers inadvertently unleashed by the technological basis of social production under capitalism. In other words, the material basis for the realization of the utopian archipelago— as theorized by Jameson—is borne of the products of socialized labor under capitalism in its global phase. Reflecting on Marx's letter to Zasulich, García Linera imagines the conditions of possibility that modern capitalist society enables for the *ayllu* to become a universal social form:

> When the value form not only seizes the conditions of production in their plenitude . . . but does so in such a way that it is the value relation itself that takes over the technical-processual order and the real intentionality of the production process, then the phase of transition has been seized by capitalism, which at the same time that it makes private property . . . the institutional measure of its material reality, it

---

48   Andreas Malm, *The Progress of This Storm: Nature and Society on a Warming World* (New York: Verso, 2018); Rivera Cusicanqui, *Un mundo ch'ixi es posible*.

49   García Linera, *Hacia el gran ayllu universal*.

likewise casts it as a burden and social curse to the forces that it has, inadvertently, awakened from their slumber. With this, *the regime of property and of capitalist labor emerges as a historical anachronism that should give way to the restoration of the archaic community, only this time with a planetary content sustained by the accomplishments of social-universalized work, involuntarily aroused throughout all these centuries of human history.* The universal community underpinned by directly universal labor, in which the laboring individual reclaims its own productive activity as something joyful, and in its originary unity with nature as the living and sacred body of human self-determination, is what we refer to the "tertiary" formation of society.[50]

According to Rivera Cusicanqui, new indigenous struggles against mining, infrastructure, and energy projects in Latin America are forging new transnational alliances, driven by the aspiration to supersede the pitfalls of localism.[51] The task ahead, she argues, is to expand these networks of solidarity even further so that they can also speak to the struggles of migrants and refugees, of stateless peoples facing oppression and persecution, and of women and children working inhumane jobs in maquilas and factories across the world.[52] These formations of sociopolitical contestation, Rivera Cusicanqui notes, evince the early stirrings of what she terms *ch'ixi* modernity: the structural combination of disparate elements that come together without being hybridized or fused—in the same way that marbled gray appears to the senses as a distinct color but is an amalgamation of black and white dots.[53]

Local communities' encounters with technological infrastructures of extraction—from power plants, to surveillance cameras to laptops and smartphones—are laying the foundations for the novel framework of generalized interdependence prefigured by notions of *ch'ixi* modernity, the universal *ayllu*, and the utopian archipelago. Although these technological systems breed socioecological degradation and dispossession, subsequent sections show that they often also breed vibrant forms of political organization that bring together city and noncity space in new

50 Ibid., 111. Emphasis added.
51 Rivera Cusicanqui, *Un mundo ch'ixi es posible*, 307–308.
52 Ibid., 308.
53 Ibid., 226.

and ever more complex ways. In other words, the irruption of these infrastructural elements across the countryside can provide the means by which the commune-form or *ayllu* becomes capable of tearing open the fetters of its insularity and expanding the framework of its societal relations. At the onset of the commodity supercycle, Latin America became the main destination for capital flows aimed at developing transcontinental infrastructural basis for resource extraction. As of 2011, the region was the most popular destination for mineral prospecting, attracting 25 percent of global capital allocations, mainly in Mexico, Chile, Peru, Brazil, Colombia, and Argentina.[54] In general, from 2003 to 2014, the region witnessed massive inflows of foreign direct investment, reaching an all-time record high of $174.5 billion in 2012 despite an overall slowdown in the global economy.[55]

According to the World Bank, much of the infrastructure investment in the region has been mobilized through "public-private partnerships," with 845 infrastructure projects developed between 2000 and 2009, accounting for an overall $310.3 billion investment.[56] With this, Latin America accounts for 30 percent of infrastructure investment in developing countries. Telecommunications, energy, and transport are the most successful sectors, attracting 47, 31, and 20 percent of regional investment, respectively.[57] As a result of these monetary flows and institutional mediations, seaports, railways, transmission lines, pipelines, roads, tailings dams, power plants, telecommunication infrastructures, and their attendant sociotechnical arrangements now stretch across the continental landscape, linking local sites of extraction with global markets. This technological and territorial reconfiguration has thoroughly transformed the countryside. To cite a few examples, by the end of the 2000s, an estimated 55 percent of Peru's highland peasant communities were being affected by mining concessions.[58] As of 2014, 65 percent of Ecuador's Amazon basin was either concessioned or available

---

54   Martín Arboleda, "Spaces of Extraction, Metropolitan Explosions: Planetary Urbanization and the Commodity Boom in Latin America," *International Journal of Urban and Regional Research* 40, no. 1 (2016): 96–112.

55   ECLAC, *La Inversión Extranjera Directa en América Latina y el Caribe* (Santiago: Economic Commission for Latin America and the Caribbean, 2012).

56   World Bank, *Private Participation in Infrastructure in Latin America and the Caribbean in the Last Decade* (Washington, DC: World Bank, 2010).

57   World Bank, *Private Participation in Infrastructure.*

58   Bebbington et al., "Anatomies of Conflict."

for concession for oil extraction, and 55 percent of Bolivia's surface was considered available for mineral forecasting.[59]

As chapter 3 illustrated, the mining industry's tendency to become more capital-intensive and interconnected with the global economy has evolved alongside the casualization and degradation of labor. However, it has also aroused mounting levels of social revolt and political contestation. Indeed, as several studies have concluded, there is a close relationship between the intensification of extraction and social mobilization in Latin America, with increasing numbers of communities and social movements opposing mining, agribusiness, logging, energy, and oil extraction projects.[60] New forms of solidarity between local communities and international advocacy networks have emerged, linking local communities and large urban agglomerations in mutually transformative ways. Organizations such as MiningWatch Canada and London Mining Network in the global North and the Latin American Observatory for Environmental Conflicts (OLCA in Spanish) and the Observatory for Mining Conflicts in Latin America (OCMAL) in the global South, have developed strong and densely inter-woven networks of cooperation and political solidarity with hundreds of communities opposing investment projects on the ground.

The unfolding of technological infrastructures for large-scale primary-commodity production has therefore ushered in a revival of communes and of the ancient *ayllu* across Latin America, including in urban areas and megacities.[61] Composed mainly of indigenous peoples,

---

59   Ibid., 245.

60   Anthony Bebbington, Denise Humphreys Bebbington, Jeffrey Bury, Jeannet Lingan, Juan Pablo Muñoz, and Martin Scurrah, "Mining and Social Movements: Struggles over Livelihood and Rural Territorial Development in the Andes," *World Development* 36, no. 12 (2008): 2888–2905; Maristella Svampa and Mirta Antonelli (eds.), *Minería transnacional, narrativas del desarrollo y resistencias sociales* (Buenos Aires: Editorial Biblos, 2009); Gudynas, "Agropecuaria y nuevo extractivismo"; Bebbington, "Underground Political Ecologies"; CINEP (Centro de Investigación y Educación Popular), *Minería, conflictos sociales y violación de derechos humanos en Colombia* (Bogotá: CINEP, 2012); Padilla, "Minería y conflictos sociales"; ECLAC, *Desarrollo minero y conflictos socioambientales: los casos de Colombia, México y Perú* (Economic Commission for Latin America and the Caribbean, 2013); Anthony Bebbington and Jeffrey Bury (eds.), *Subterranean Struggles: New Dynamics of Mining, Oil, and Gas in Latin America* (Austin: University of Texas Press, 2013).

61   Svampa and Álvarez, "Modelo minero"; Raúl Zibechi, *Territories in Resistance: A Cartography of Latin American Social Movements)* Oakland, CA: AK Press, 2012); García Linera, *Hacia el gran ayllu universal*; Gago, *La razón neoliberal*; Webber, "Indigenous Community"; Ulloa, "Feminismos territoriales en América Latina."

peasants, women's movements, and migrants, among other "motley" fragments of the subaltern and laboring classes, these communal forms have emerged with the aspiration to defend livelihoods and ways of living from the disruptive socioecological effects of capital-intensive raw-materials production. More than being merely oppositional, these struggles are concerned with inventing new territorial configurations in which value relations are overthrown and replaced by renewed modes of sociality.[62]

Despite the diversity of the aforementioned movements, Zibechi outlines some of their most recurrent features.[63] First, they are concerned with appropriating and repurposing space: squatting, occupations, and blockades are some of the tactics employed to recuperate land and remodel sociospatial relations. Second, these struggles seek autonomy from the state and traditional political parties. Third, they are premised on the assertion and revalorization of ethnic, popular, and/or gendered identities that are radically at odds with those of capitalist heteropatriarchy. Fourth, these communes are "pedagogical projects" in their own right because they frequently involve training their own organic intellectuals and producing their own vernacular scientific knowledge about the world. Fifth, women tend to play important roles as organizers, and the concept of the family is expanded beyond private spaces to become mobilized both as a polity and productive unit. Finally, these emerging forms of territorial politics are generally concerned with reinventing relations of production beyond the vertical and hierarchical nature of the Taylorist organizational form.[64]

In her account of recent social mobilization in Bolivia, Raquel Gutiérrez notes how indigenous peoples overcame ethnic and social divisions by a collective recovery of communal and decommodified forms of social mediation.[65] Thousands of communards in El Alto, Cochabamba, and Chapare transformed their neighborhoods into public assemblies where new mechanisms of social reproduction and of

62 Marisol de la Cadena, "Indigenous Cosmopolitics in the Andes: Conceptual Reflections beyond 'Politics,'" *Cultural Anthropology* 25, no. 2 (2010): 334–70; Gudynas, "Value, Growth, Development"; Escobar, *Designs for the Pluriverse*.

63 Zibechi, *Territories in Resistance*.

64 Ibid.

65 Raquel Gutiérrez, *Los ritmos del pachakuti: Movilización y levantamiento popular-indígena en Bolivia (2000–2005)* (Buenos Aires: Tinta Limón, 2008).

public deliberation were asserted as a political weapon against the extractive/rentier forms of neoliberal governance imposed by the Bolivian state before the Evo Morales administration.[66] These emerging forms of political organization are not exclusive to Bolivia and are becoming a common feature of the political geography of Latin America in general, where the expansion of commodity frontiers poses a direct threat to the lives and means of subsistence of peasant and indigenous communities. Because some of these popular mobilizations against extraction are often led by women, Ulloa refers to them as "territorial feminisms." Some of the most dynamic new expressions of political mobilization against extraction in Latin America, she writes, enact "a feminist perspective of space, one which posits another geopolitics, an alter-geopolitics, as well as alternative territorial visions and processes of care across multiple scales, starting from the body-territory."[67]

Labor insurgency in the Latin American mining supply chain has tended to mirror in certain respects the modalities of horizontal, networked, and decentralized mobilization prevalent among subaltern movements. In sharp contradistinction to the corporatist and institutionalized labor politics that defined the national-developmentalist era in Latin America, García Linera explains that outsourcing, labor flexibilization, and the horizontal integration of the mining industry have enabled a reconstitution of class consciousness that no longer coalesces around the workplace but around territorial systems such as neighborhood councils, skill-trade associations, and so forth.[68] A new configuration of proletarianization began to emerge with the computerization of social production after the 1990s—one which, according to García Linera, "includes more workers than that of decades ago, but is physically fragmented into tiny shops ... with precarious, temporary contracts, systems of promotion based on competition and unions lacking legitimacy in the eyes of the state."[69] One of the landmark events that crystallized this particular shift in the new composition of proletarian politics was the July 2006 occupation of the Escondida mine in Chile. In the face of rampant outsourcing, subcontracting, and wage repression,

---

66 Gutiérrez, *Los ritmos del pachakuti*; Zibechi, *Territories in Resistance*.
67 Ulloa, "Feminismos territoriales en América Latina," 126.
68 García Linera, *Hacia el gran ayllu universal*.
69 Ibid., 213.

the workers of Escondida did away with old corporatist practices and occupied mining facilities and the roads that connected them with ports and neighboring cities.[70]

Echoing tactics first introduced by the student protests earlier that year, and perhaps foreshadowing the Occupy Wall Street movement, the unionized workers of Escondida set up tents inside the mine, in association with a women's movement composed of the mothers, wives, partners, and daughters of the workers. The makeshift camp lasted for twenty-five days, in which the workers played music, engaged in activities of radical pedagogy, and established horizontal and communitarian modes of social life.[71] From that moment on, new varieties of labor organization began to emerge in Chile, connected both to the mining companies and their expanding constellation of contractors. The social composition of these new trade unions is markedly different from those of the past because it has come to include all those "plebeian" components of the laboring classes (migrants, women, temporary workers, semiproletarianized peasants) traditionally deemed external to "proper" labor politics.

In November 2014, the Mining Federation of the North (Feminort) was created with the purpose of bringing together subcontracted workers of the Collahuasi mine. Initiatives of this sort, which have been aimed at the organization of temporary workers, reached its latest iteration with the creation of the Frente de Trabajadores Nelson Quichillao in 2015. Named after the worker murdered by the police during the El Salvador mine occupation earlier that year, this organization has been actively seeking the implementation of a "master agreement" with Chile's major mining firms to improve the conditions of the temporary and flexible workforces of the mining supply chain. As the next section points out, the context of Asian industrialization displays very similar characteristics. This, it will be argued, warrants a theory of revolutionary subjectivity that is attentive to cultural specificity while also drawing broader connections between laboring subjects that, on the surface, appear unrelated.

---

70   Orlando Caputo and Graciela Galarce, "La huelga en Minera Escondida y la reactivación del movimiento social en Chile," OSAL-Observatorio Social de América Latina 7, no. 20 (2016): 117–27.

71   Caputo and Galarce, "La huelga en Minera Escondida."

Labor Unrest in Asia and the Social Determinations of Revolutionary Subjectivity

Although struggles against "extractivism" have tended to be framed as culturally and politically specific to the realities of "resource-rich" nations, a closer look at advanced manufacturing reveals significant resemblances. To begin with, the industrial proletariats of East Asian economies have been largely drawn from the rapidly shifting country-sides, and important segments of the workforce continue to be linked in certain ways to artisanal farming and communal modes of sociality. The case of China is particularly illustrative. Market reforms implemented by the Communist Party in the 1980s heavily undermined subsistence farming and peasants were besieged by debt and escalating taxes. This created a mass exodus toward the cities, especially in the onset of the newly formed special economic zones (SEZ), when peasants became temporary workers in assembly lines, construction, and cash crop agriculture.[72]

Ever since the establishment of the *hukou* system, workers registered in a "rural" district have not been allowed to settle permanently in "urban" districts. The main effect, Cheng points out, has been a highly precarious and flexible workforce of semiproletarianized peasants.[73] According to Friedman, the most important section of the Chinese proletariat is composed of precarious migrant workers, and this indicates an overall a race to the bottom in terms of working-class living standards.

Although Chinese workers—and Asian workers generally—are typically portrayed as docile victims of capitalist industrialization, China has become a breeding ground for an important new cycle of labor unrest, as well as for the invention of new forms of labor organization.[74] Although worker insurgency during the 1990s was fragmented and

---

72 Cheng, "Hukou"; Ling, "Survival and Collective Struggles."

73 Cheng, "Hukou."

74 Beverly Silver and Lu Zhang, "China as an Emerging Epicenter of World Labor Unrest," in Hung (ed.), *China and the Transformation of Global Capitalism*; Minnie Chan, "Strike," in Al (ed.), *Factory Towns of South China*; Ho-Fung Hung, *Protest with Chinese Characteristics: Demonstrations, Riots, and Petitions in the Mid-Qing Dynasty* (New York: Columbia University Press, 2013); Friedman, *Insurgency Trap*; Ren et al. (eds.), *China on Strike.*

articulated around individual shop-floor grievances, the first years of the twenty-first century witnessed a shift in the intensity and scale of such struggles. According to official figures from the Chinese government, mass protest increased from 10,000 incidents involving 730,000 protestors, in 1993, to 60,000 incidents involving more than 3 million protestors in 2003.[75] Between 2004 and 2006, a concentrated surge of strikes cropped up in large, capital-intensive, and foreign-funded electronics factories.

Due to the enlargement of its industrial basis, China's infrastructural matrix has undergone a dramatic reconfiguration. This has become a central source of political dispute and social instability, unsurprising given the sheer scale of the transformations. In 2005, China had 41,000 kilometers of highways, and in just nine years this had multiplied by two and a half and the country now boasts the most extensive network in the world. Before 2003, China had no working high-speed rail lines; now it has 12,000 kilometers, the most extensive in the world, and is expected to expand to 50,000 kilometers by 2020. As of 2015, eighty-two new civil airports were being built and 101 airports were being overhauled and expanded. One, in the south of Beijing, would cost around $14 billion and would displace 116,000 people.[76] In light of this, Arrighi argues that some of the most incendiary issues in contemporary waves of social revolt in China have been the diversion of land from small-scale farming to capital-intensive uses, environmental degradation, changes in land use due to infrastructural modernization, and the corruption of local government officials—grievances that are highly reminiscent of those against resource extraction in Latin America.[77]

In a manner of labor organizing that resembles the Latin American "territorial feminisms" described by Ulloa,[78] one of the most characteristic features of recent waves of labor insurgency is that they have been led by female workers.[79] Female workers mobilized to protect union leaders and their male colleagues from management and developed tactics to shield protestors from police repression at crucial times of industrial

---

75   Silver and Zhang, "China as an Emerging Epicenter," 175.
76   Shepard, *Ghost Cities of China*, 6.
77   Arrighi, *Adam Smith in Beijing*, 377.
78   Ulloa, "Feminismos territoriales en América Latina."
79   Ling, "Survival and Collective Struggles."

action.[80] For example, "male workers were more likely to be imprisoned when confronting the police, while female workers were relatively safe given that the police would not dare touch them. (Female workers would scream 'Harassment!' if they did)."[81] For this reason, they devised a way to arrange women in the front and outside to protect men in the center.[82] As in resource extraction, these new varieties of labor insurgency have been unorganized, spontaneous, and typically rely on methods such as roadblocks, demonstrations, occupations, and collective petitions.[83]

China's 2008 labor reforms mirror those being put in place in Latin America's primary sector, which enable employers to use open-ended work contracts to push outsourcing, subemployment, and diverse modalities of subcontracting.[84] Between the implementation of the Labor Contract Law in early 2008 and late 2010, the number of subcontracted workers in China leaped from 20 million to 60 million.[85] Far from stifling labor insurgency, this reform galvanized new configurations of class consciousness that extended worker mobilization beyond capital-intensive factories to include laborers in other sectors of the economy. As a result, worker unrest is no longer found exclusively in the "hotbeds" of China's large-scale manufacturing like the Pearl and Yangtze River Deltas but has moved to encompass other areas and types of workers, such as bus drivers, street cleaners, retail workers, and even some segments of white-collar workers.[86] Although the demands of steady laborers are more defensive in nature, migrant and subcontracted workers pose the most *offensive* demands.[87] This leads Friedman to the conclusion that migrants represent the future of the working class—in both economic and political terms. Far from being a tendency exclusive to China, Lüthje and coauthors' study of the electronics industry concludes that the workforce of East Asian megafactories is overwhelmingly made up of low-paid, racialized migrants, mostly women.[88]

---

80 Ling, "Survival and Collective Struggles"; Ren et al. (eds.), *China on Strike*, chapter 4.
81 Ren et al. (eds.), *China on Strike*, 70.
82 Ibid.
83 Ibid.
84 Friedman, *Insurgency Trap*.
85 Ibid., 8.
86 Ren et al. (eds.), *China on Strike*.
87 Friedman, *Insurgency Trap*.
88 Lüthje, *From Silicon Valley to Shenzhen*.

The laboring subjects that lie at both ends of the mining supply chain, therefore, have important characteristics in common. That these new waves of labor and popular revolt are driven mainly by "plebeian" components of the working class—in both Asia and Latin America— indicates how deeply rooted Mbembe's idea of the "becoming black of the world" is in the antagonism between capital and labor.[89] But the workforces at both ends of the infrastructural systems of extraction have a further relevant element in common. According to Ho-fung Hung, most varieties of social protest in contemporary China echo those of mid-Qing China of the eighteenth and nineteenth centuries, especially as they tend to be framed in the idiom of "livelihoods" and "subsistence rights"—as is also the case in Latin America, where subaltern struggles are articulated around questions of territoriality for the same reasons.[90] As opposed to Western conceptions of rights, which emphasize individual freedom, Chinese political advocacy is cast through a Confucianist schema of collective subsistence and moral economies.

By exploring the repertoires of sociopolitical contestation unfolding among the Chinese laboring classes, Hung challenges the teleology of history of traditional political economy, demonstrating that China has not followed a linear path of political evolution whereby non-Western tactics are superseded by those more directly anchored in the European Enlightenment tradition.[91] However, this does not mean that the popular culture of the Chinese laboring classes has not been influenced or shaped in any way by Western traditions. For Hung, there have been important cultural hybridizations in China's language of political advocacy, but Western canons have changed the Chinese cultural reservoir by *enriching* it rather than by *replacing* it.[92] A similar trend can be identified across other rapidly industrializing countries of East Asia, where tactics of protest and labor insurgency remain to some extent anchored in communal mechanisms of deliberation typical of the local peasant culture. The Gwangju Commune of South Korea in 1980, as well as the protests that occurred at Bangkok's

---

89   Mbembe, *Crítica de la razón negra*.

90   Hung, *Protest with Chinese Characteristics*; Maristella Svampa, "Conenso de los commodities, giro ecoterritorial, y pensamiento crítico en América Latina," *Revista OSAL* 32 (2012): 15–38; Zibechi, *Territories in Resistance*.

91   Hung, *Protest with Chinese Characteristics*.

92   Ibid.

Thammasat University in 1973; in Beijing's Tiananmen Square in 1989; in Patan, Nepal, in 1990; and in Taipei's Chiang Kai-shek Square in 1990 are relevant examples.[93]

It would appear that the evolving forms of revolutionary consciousness among these communities (in both Asia and Latin America) have as their foundation the precapitalist, culturally specific, and noncommodified elements of individuals. A closer look at the evidence and at the nature of these struggles, however, reveals a more complex process at work. The question of revolutionary subjectivity has arguably sparked the most discrepancies between the traditions of form-analysis Marxism that inform this book. Although most of these traditions converge on the aspiration to discover the social constitution of economic categories (i.e., capital, labor, money, state, forces of production) as perverted or alienated modes of existence of human sensuous practice, they disagree on the nature of the agent whose revolutionary action is to abolish the material reproduction of such categories. Debates over the latter aspect span several academic publications and are widely documented in the specialized literature, so a detailed discussion is beyond the scope of this book.[94] However, exploring some of the core positions in the debate can be useful to elucidate the foundation of the emerging forms of revolt at the heart of the geography of extraction, as well as whatever transformative potential they may contain. Despite the internal nuances in each of the positions articulated in these Marxological debates, the crux of the disagreement stems from whether the grounds of revolutionary praxis are to be found in an "uncontaminated" human essence or "residue" external to the cash nexus, or if it instead arises from individuals in their status as personifications or *dramatis personae* of capital—that is, from

---

93   George Katsiaficas, "The Commune: Evolving Form of Freedom," *ROAR* 1 (2016): 13–21.

94   Guido Starosta, "Editorial Introduction: Rethinking Marx's Social Theory," *Historical Materialism* 12, no. 3 (2004): 43–52; Starosta, *Marx's Capital*; Starosta, "Fetishism and Revolution"; Werner Bonefeld, "On Postone's Courageous but Unsuccessful Attempt to Banish the Class Antagonism from the Critique of Political Economy," *Historical Materialism* 12, no. 3 (2004): 103–24; Bonefeld, "Emancipatory Praxis and Conceptuality in Adorno," in John Holloway, Fernando Matamoros, and Sergio Tischler (eds.), *Negativity and Revolution: Adorno and Political Activism* (London: Pluto Press, 2008); Postone, "Critique and Historical Transformation"; Axel Kicillof and Guido Starosta, "Value Form and Class Struggle: A Critique of the Autonomist Theory of Value," *Capital and Class* 92 (2007): 13–40.

the "laws of motion" of capital as the alienated subject of the process of social metabolism.

For the Open Marxist tradition, the grounds of political action are not located in commodity-determined practice itself but in the negated human content that eludes capitalist commodification and exists in an undistorted form in a prerevolutionary situation.[95] As Bonefeld puts it, "Essence [*Wesen*] exists in the mode of being denied—as mischief [*Unwesen*]. That is to say, sensuous human practice subsists against itself in the form of, say, freedom as wage slavery."[96] Put differently, Bonefeld considers that this precommodified essence cannot be fully obliterated but objectively exists in the inverted world of capital. In his words, "sensuous being exists within the concept of variable capital in the mode of being denied."[97] Holloway's more recent work espouses a similar approach, proclaiming that freedom already exists in an undistorted form in the context of commodification.[98] As Holloway puts it, freedom exists "in, against, and beyond capital."[99] This interpretation of the conditions of transformative political action is succinctly captured by Bonefeld's dictum that "variable capital does not go on strike. Workers do."[100] The tendency to posit an irreducible human essence as the foundation of transformative political action is not exclusive to Open Marxist authors, but is quite possibly one of the most predominant within the academic left broadly considered.

Henri Lefebvre's ideas on the politics of space and on everyday life—which have inspired decades of scholarship in urban studies and geography—strongly resonate with the Open Marxist reading of political action described above.[101] For Lefebvre, there is a residuum of human subjectivity and style that capital has been unable to subsume, subvert, or control, and it is from such residuum that revolutionary practice is to

---

95   For an overview, see Starosta, "Fetishism and Revolution."

96   Bonefeld, "Emancipatory Praxis," 136.

97   Ibid., 139.

98   John Holloway, "Stop Making Capitalism," in Bonefeld and Psychopedis (eds.), *Human Dignity*; John Holloway, *Crack Capitalism* (London: Pluto Press, 2010); Richard Gunn and Adrian Wilding, "Holloway, La Boétie, Hegel," *Journal of Classical Sociology* 12, no. 2 (2012): 173–90.

99   Cited in Gunn and Wilding, "Holloway, La Boétie, Hegel," 174.

100   Cited in Starosta, "Fetishism and Revolution," 376.

101   Charnock, "Challenging New State Spatialities."

emerge.[102] Everyday life, according to Lefebvre, is where such "irreducibles" are therefore to be found. Besides authors in the Marxist tradition, the sort of pastoralism that ontologizes the class struggle, rooting the nature of transformative action into elements that transcend the material metabolism of social life (i.e., an uncorrupted human essence, cultural attributes, or ahistorical moral imperatives) is also particularly predominant in studies of extraction and of indigenous politics. Thus the intrinsic virtue or other specific "residue" of the subaltern subject, existing prior to the context of commodification, is deemed the genuine basis of political agency.

By contrast, Moishe Postone's idiosyncratic reinterpretation of Marx's critique of political economy posits that the task of critical theory is to discover emancipatory consciousness as *socially constituted* by the directionally dynamic movement of capital itself.[103] It is not that social groupings organized around religious, ethnic, national, or gender issues play no important roles historically and politically. Rather, his reading of the basis of social antagonism points toward the fact that such expressions of political conflict are embedded within a framework of social interdependence that renders them moments of a dynamic totality, thus becoming dynamic and totalized in their conflict.[104] With the historical emergence of the commodity as a totalizing social form, Postone claims, a mode of social mediation comes into being that is abstract, homogeneous, and premised on universal material connection between individuals. It is in the unfolding of such alienated modes of universality—not in transhistorical or teleological evolutionary processes—where, according to Postone, the origins of emancipatory consciousness should be sought. In a similar vein, Donna Haraway's concept of the cyborg also grasps the postfoundationalist thrust at the heart of Marx's understanding of technological change in capitalist society.[105] For Haraway, the cyborg is at once metaphor and object of the developmental potentialities that capitalist technologies inadvertently bring to the "plebeian" elements of the working class. Far from emerging from the myth of an

---

102   Greig Charnock and Ramón Ribera-Fumaz, "A New Space for Knowledge and People? Henri Lefebvre, Representations of Space, and the Production of 22@Barcelona," *Environment and Planning D: Society and Space* 29 (2011): 613–29.

103   Postone, *Time, Labor, and Social Domination.*

104   Ibid., 320.

105   Haraway, *Simians, Cyborgs and Women.*

original unity or an identification with pristine nature, the cyborg is the "illegitimate offspring of military and patriarchal capitalism" that, *qua* oppositional subject, is unfaithful to its origins.[106]

The significance of Postone's critical-theoretical framework, in my view, stems from taking seriously Marx's insistence that the material transformation of social forms of labor is where the key to the abolition of capital resides. One of the most contentious aspects of Postone's presentation, however, is his view that the proletariat cannot assert itself as a revolutionary subject because it is a personification or *mode of existence* of capital and thus can only be the bearer of alienated practice.[107] Although Postone's critics are right to point out that such an assumption is problematic, his work provides the fundamental elements of a theory of political agency based on the material transformation of the human life process, not on first principles or moralizing discourses, as is usually the case.[108] Building upon this element of Postone's approach and on a rigorous reconstruction of Marx's mature work, Guido Starosta and Juan Iñigo Carrera have developed a theory of revolutionary subjectivity that posits transformative political action as eminently grounded in commodity-determined practice, not in any residual substance or abstract material content deprived of social determinations.[109] Any power individuals might have to radically transform the world, Starosta argues, must be a concrete form of the commodity itself. When the workers organize the revolutionary abolition of the capitalist mode of production, they do so as personifications of the inverted existence of the powers of their social labor—that is, capital.[110] In Starosta's words, this means that:

---

106   Ibid.

107   Marx's treatment of the labor process in modern society, according to Postone, supports the claim that the overcoming of capitalism would not entail the self-realization of the proletariat. For Postone, then, the logic of Marx's presentation does not support the notion that the proletariat is the revolutionary subject. The reappropriation of the vital capacities seized by capital can only be possible if the historical process of alienation—which includes proletarian labor—is abolished.

108   For a critique, see Starosta, "Rethinking Marx's Social Theory"; Bonefeld, "On Postone's Courageous but Unsuccessful Attempt."

109   Starosta, "Fetishism and Revolution."

110   Iñigo Carrera, *El Capital*; Starosta, *Marx's Capital*.

revolutionary powers are not "self-developed" by the workers, but are an alienated attribute that capital puts into their own hands through the transformations of their productive subjectivity produced by the alienated socialisation and universalisation of labour through which the production of relative surplus value takes place. This is the reason why *revolutionary consciousness is itself a concrete form of the aliena-tion of human powers as capital's powers.*[111]

Technology plays a definitive role in the process whereby capital meta-morphoses into conscious revolutionary action. The implications of this claim are more readily and blatantly observed in the industrialization of the countryside. The "conscious, technological application of science," says Marx, is what replaces the traditional way of exerting production in the countryside and at the same time completes the disintegration of the "primitive familial union which bound agriculture and manufacture together when they were both at an undeveloped and childlike state." Therefore, when the peasant undergoes such radical metamorphosis, Marx notes how class antagonisms and the concomitant need for social transformation tend to acquire the same intensity in the country as in the city.[112] Evidence from China's recent industrialization supports the claim that rising capital-intensity in the labor process is correlated with height-ened labor insurgency and sociopolitical contestation.[113] As Silver and Zhang point out, a new and militant working class has emerged in the onset of the post–Deng Xiaoping era in China, as capital moves to encom-pass ever larger swathes of the country.[114] Although worker unrest is more visible in cities, the expansion of mechanized systems for capitalist production throughout the country's erstwhile rural regions has proceeded apace with the emergence of new sites of labor insurgency. As we will see in the following section, a similar development is taking place at the other side of the Pacific Ocean, where copper and other metals are first intro-duced into the vast circulatory system of the planetary mine.

---

111   Starosta, "Fetishism and Revolution," 391–92. Emphasis in original.
112   Marx, *Capital,* Vol. 1, 637.
113   Chan, "Strike"; Silver and Zhang, "China as an Emerging Epicenter"; Friedman, *Insurgency Trap*; Ren et al. (eds.), *China on Strike.*
114   Silver and Zhang, "China as an Emerging Epicenter."

Struggles against Large-Scale Mining in the Huasco Valley

With rebellion, García Linera writes that the *ayllu* ceases to be a relic of ancient times and "presents itself anew as a rational foundation for a superior form of autonomously producing social life."[115] In the Huasco Valley, the introduction of technologically upgraded industrial systems for resource extraction has created the material conditions for the community to explode its own boundaries and project its radical political praxis across scales. Alongside job destruction and instability, the computerization and mechanization of resource extraction has also underpinned renewed processes of knowledge production in formal settings as well as in the course of workers' practical interaction with technological infrastructures. The arrival of mining to the valley triggered an explosion of tertiary education programs for people providing services to mining companies.[116] According to official data from the Chilean Internal Revenue Agency (SII), revenues obtained by local education centers increased almost eightfold from 2005 to 2012, side by side with an equally sharp increase in the fixed capital expenditures of the largest mining projects in the region.[117] Between 2001 and 2008, several new technical and tertiary degrees were created, including computer analysis and programming, legal services, instrumentation and automation, and industrial equipment maintenance.[118]

The modernizing drive of transnational capital in the sphere of reproduction has also been fundamental for the emergence of new configurations of human productive subjectivity and new productive attributes because, as a local activist and agricultural worker pointed out, satellite television, cellular phones, computers, and internet for everyday use have followed the arrival of mining and energy corporations to the Huasco Valley.[119] Also, as explored in the previous chapter, retail electronics stores and car dealers populate the streets of Vallenar. To

---

115   García Linera, *Plebeian Power*, 156.

116   Martín Arboleda, *Resource Extraction and the Planetary Extension of the Urban Form: Understanding Sociospatial Transformation in the Huasco Valley, Chile*, PhD thesis presented to the University of Manchester, UK, 2015.

117   Available at the SII's website (sii.cl).

118   Martín Arboleda, *Resource Extraction*.

119   Interview with a member of Consejo de Defensa del Valle del Huasco, November 28, 2013.

mitigate resistance to large-scale mining, a local activist notes that Barrick Gold gave laptop computers to the local community and installed free wireless internet in Alto del Carmen.[120] As materialist interpretations of technological change point out, the introduction of large-scale systems of machinery to the labor process have tended to exert the dual effect of deskilling the workforce and developing the scientific consciousness of the collective laborer.[121] Briefly put, the materialist worldview reveals how the same historical determinations that revolutionize instruments of production also mediate and expand the material powers of the individual *qua* productive subject.

Ever since the Pascua Lama gold mine opened in 2001, social mobilization has been fierce and wide-ranging. The people of the Huasco Valley have been developing strong, politically organized communities and advocacy networks to stop the encroaching threat that mining and energy activities represent to their territories and ways of living. What is important to note here is that these apparently free-floating social mobilizations are based on the revolution in material conditions of production. According to Lukács, to the extent that the productive process expands and becomes more complex, the capitalist must set into motion even more elaborate calculation and rationalization dynamics.[122] For the working classes, on the other hand, this development has a different class meaning because it underlies the "abolition of the isolated individual": it leads workers to become conscious of the social character of labor and its intrinsically revolutionary potential.[123] According to my interviewees, before the arrival of transnational mining there was no sense of community within the villages or between them. Now, although they have considerable disagreements regarding mobilization strategies, demands, and worldviews, locals nonetheless feel themselves part of a growing community that stretches throughout the Huasco Valley and northern Chile.

These outsourced, precarious, and often racialized laboring subjects have come to constitute the backbone of social resistance against mining. At the local level, community organizations for political mobilization

---

120   Interview with a member of Creando Valle, December 1, 2013.
121   Starosta, *Marx's Capital*; Postone, *Time, Labor, and Social Domination*.
122   Lukács, *History and Class Consciousness*, 171.
123   Ibid.

have flourished, the most relevant being the Pastoral Salvaguarda de la
Creación, the Council for the Defense of the Huasco Valley, Brigada
SOS Huasco, Comunidad Agrícola Diaguita los Huascoaltinos,
Asamblea por el Agua del Guasco Alto, Comité Ecológico y Cultural
Esperanza de Vida, Unidos por el Agua, Comunidad Diaguita los
Tambos, Comunidad Diaguita Patay Co, Comunidad Diaguita Montañas
Fértiles, Asociación de Pequeños Agricultores de San Félix, Junta de
Vecinos Piedras Juntas, Junta de Vecinos Chollay, Pajareteros Alto del
Carmen, and the Huasco Valley Socioenvironmental Movement
(Movimiento Socioambiental del Valle del Huasco). These organiza-
tions span the communities of Vallenar, Freirina, Huasco, and Alto del
Carmen, the four villages located within the Huasco Valley.

It is possibly the radical transformations of labor engendered by the
geographical expansion of capital-intensive systems of primary-
commodity production that have instilled in peasant communities an
increasing awareness of their own material conditions of existence as an
alienated yet revolutionary social subject. As noted in chapter 3, such
mining technologies have become a source of ecological destruction,
proletarianization, labor instability, and precariousness. In volume 1 of
Capital, Marx observed how the sphere of agriculture was where industry had the most revolutionary effects because it annihilated the "peasant," replacing it with the "wage-labourer."[124] Indeed, as the peasantry
becomes communicative and active, Hardt and Negri suggest, it ceases
to exist as a separate political category, ultimately exploding the boundary between city and country. The figure of the peasant thus emerges
from its passive and isolated state "like the butterfly emerging from its
chrysalis."[125] In so doing, peasants discover themselves as one among
other figures of labor who, *despite cultural difference*, share common
conditions of existence. Paradoxically, then, Hardt and Negri argue that
the final victory of the peasant revolution is the end of the peasantry—
its own destruction as a class.

Because labor markets have become mobile and flexible, the pervasive
territorial effects of capital-intensive primary-commodity production in
the Huasco Valley have not been restricted to situated areas of production

---

124   Marx, *Capital*, Vol. 1, 637.
125   Michael Hardt and Antonio Negri, *Multitude: War and Democracy in the Age
of Empire* (New York: Penguin, 2004), 124.

but have spilled over to the whole geography of the region. As a result, political organization and its concomitant expressions of rebellion and antagonism have extended to agroindustrial complexes, mines, and energy firms across the rapidly urbanizing landscapes of northern Chile. For this reason emancipatory subjectivity—and thus the seeds for the supersession of capitalism—needs to be understood as immanent in the very unfolding of the reified forms of social mediation of modern society. The consolidation of the emancipatory social subject that has followed the territorial configurations of the commodity supercycle has therefore not resulted from an ahistorical or transcendental moral imperative—as most variants of leftist thought would be inclined to suggest. It is rather the most genuine product of the revolutionary transformations in the technical composition of labor that have followed the sprawling networks of material intercourse enabled by contemporary technologies of extraction.

In terms of strategy, social mobilization activities against mining and energy megaprojects are diverse, with the internet and social media important tools for disseminating blog entries, documentaries, and statements as well as for creating international alliances and organizing events.[126] Several forms of direct action, such as demonstrations, marches, occupations, blockades, sabotage of infrastructures, and picket lines, have also been very effective.[127] Marches and demonstrations have also been organized in Santiago de Chile to pressure the national government and give visibility to the struggle at the national and international levels. Members of local communities have traveled on several occasions to Toronto to attend the annual general meetings (AGMs) of large mining corporations that are developing projects in the valley. There, community leaders usually denounce before shareholders the bad practices of the companies at extraction sites and in corporate governance.

Litigation has been one of the most relevant institutional strategies adopted by these social movements, which have advanced lawsuits, allegations, legal resources, and several other sorts of juridical mechanisms before national courts and international organizations such as the

---

126   Urkidi, "Movimientos anti-mineros"; interview with a member of Huasco Valley Socioenvironmental Movement, December 5, 2013; interview with a member of OLCA, December 11, 2013.

127   Interview with a member of SOS Huasco, December 4, 2013.

Inter-American Court of Human Rights. These legal mechanisms are usually aimed at denouncing flawed environmental impact assessments, unlawful degradation of the environment, noncompliance with Convention 169 of the International Labor Organization, displacement, and loss of livelihoods.[128] At the national level, litigation has been a very effective mechanism against investment projects, even to the point of making Chile drop four positions (since 2011) in the 2015 Fraser Institute's ranking for best countries for mining investment.[129] According to a member of the Fraser Institute, Chile's downgrading can be attributed to the implementation of more stringent regulatory measures as well as to a sense of "uncertainty" regarding land-tenure schemes among investors— since mining projects tend to generate conflicts with local communities.[130] According to OLCA, as of December 2013, $37 billion on energy and mining investments had been put on hiatus in Chile alone as a result of judiciary measures.[131] Due to the effectiveness of these strategies, state crackdowns on political activism are now rampant. Increased police repression and surveillance are common across the region, as well as the enactment of criminal laws to intimidate and dissuade further blockades.[132]

As a result of these strategies of sociopolitical contestation, the communities of the Huasco Valley have achieved the unimaginable,

128   Convention 169 of 1991 is a legally binding international instrument dealing specifically with the rights of indigenous and tribal peoples. Among the special measures adopted to protect the environs, livelihoods, beliefs, and ways of living of indigenous peoples, is the mechanism of consultation: Convention 169 requires that indigenous peoples are consulted on any issues that affect them, and this naturally includes the development of investment projects (such as mines, transmission lines, and power plants) in their territories.

129   mch.cl/2015/02/25/chile-ocupa-el-puesto-13-en-ranking-mundial-de-atractivo-para-inversion-minera (accessed 26 February 2015).

130   mch.cl/2015/02/25/chile-ocupa-el-puesto-13-en-ranking-mundial-de-atractivo-para-inversion-minera (accessed 26 February 2015).

131   OLCA, "Proyectos mineros y eléctricos por US$ 37 mil millones están frenados por judicialización," December 2, 2013, olca.cl/articulo/nota.php?id=103832.

132   Eduardo Mella, "El Estado chileno contra la protesta social, 2000–2010," SudHistoria 4 (2012): 73–92; Jorge Ceja Martínez, "Seguridad ciudadana, militarización y criminalización de las disidencias en México," Espacio Abierto Cuaderno Venezolano de Sociología 22, no. 3 (2013): 681–99; Kate Swanson, "Zero Tolerance in Latin America: Punitive Paradox in Urban Mobilities," Urban Geography 34, no. 7 (2013): 972–88; José Saldana, "Criminalización de la protesta y el consenso represivo," Parthenon, 2015, parthenon.pe/columnistas/jose-saldana-cuba/criminalizacion-de-la-protesta-y-el-consenso-represivo.

jeopardizing the infrastructural systems that connect extraction sites with the Pacific Ocean. With all odds against them, especially considering the manifold layers of state spatial planning and engineering, the pressures exerted by international commodity markets, and the enactment of institutional and legal frameworks aimed at attracting billions of dollars' worth of foreign direct investment (FDI), they have been able to stop the unrelenting expansion of primary-commodity frontiers. Pascua Lama, the eleventh largest undeveloped open-cast mining project in the world, at the hands of an equally major company like Barrick Gold (see chapter 6), has suffered all sorts of setbacks due to social resistance and is currently on the verge of financial and technical unviability. Domestic authorities have imposed fines ranging from $12 billion to $17 million for bad practices such as obstructing waterways (April 2007), unlawfully appropriating water resources (February 2011), degrading glaciers (January 2013), and discharging wastewater to the Huasco River (February 2013), among several others.[133] Most of the time, these fines include the suspension of activities, with deeply problematic consequences for Barrick Gold's land surveying and construction operations as well as operational costs and shareholder confidence, which has undermined the performance of its shares at the Toronto Stock Exchange.[134] Currently, Pascua Lama is suspended as a result of litigation aimed at deciding whether to impose further fines and whether the company's concession can be revoked on the grounds of bad practices and socioenvironmental unviability. Due to these obstacles imposed on its operation, Barrick Gold revealed in 2017 that it is considering reframing Pascua Lama as an underground mine—rather than open pit, as initially proposed—to reach agreement with local communities and present a new environmental impact assessment to the regulatory agencies.[135]

The operations of a pork-processing plant in Freirina with capacity for 2.5 million animals were also brought to an indefinite halt after massive roadblocks and protests by local communities. The construction of Punta Alcalde—one of the country's largest projected thermoelectric

---

133   Information about the various fines imposed upon Barrick Gold can be found online at the website of the OLCA (olca.cl).

134   De Los Reyes, "Pascua Lama in Barrick Gold's Strategy."

135   AND Radio, "Barrick prepara la reactivación de Pascua Lama como mina subterránea," July 25, 2017, adnradio.cl.

plants—was also stopped after a 2013 ruling from the Santiago Court of Appeals in favor of local communities that denounced flaws in environmental impact assessments. Also facing suspension is El Morro, an opencast gold mine with a projected investment of $2.5 billion, owned by Goldcorp and New Gold and geographically contiguous to Pascua Lama. In November 2013, the Copiapó Court of Appeals temporarily withdrew El Morro's mining license due to socioenvironmental unviability and lack of compliance with ILO Convention 169 following a lawsuit filed by fifteen Diaguita indigenous communities based near Alto del Carmen.[136]

Opposing investment projects of up to $8 billion and *actually being able to stop them* is beyond the capacities of insular social individuals lacking in social-universal interconnectedness. The sociospatial change taking place in the Huasco Valley therefore provides a privileged vantage point to see how capitalist technology can radically expand the vital capacities of the individual in its status as a personification of the forms of abstract sociability engendered by the capitalist mode of production. The primitive stirrings of the universalized archaic community envisaged by the late Marx begin underneath the layers of political organization, strategy, litigation, self-determination, physical mobility, and campaigns to stop the devastation produced by resource extraction in the "Garden of Atacama," the last fertile valley in northern Chile.

Conclusion

Writing in the 1980s, Maria Mies was already acutely aware of the fragmenting effects of new international divisions of labor structured around relocating large-scale industry to the global South. Such transnational modalities of industrial organization, Mies argued, had divided women internationally and class-wise into producers and consumers.[137] For Mies, a truly emancipatory politics had to overcome the cultural relativism that obfuscated the material interdependencies and common conditions of existence between oppressed groups worldwide.[138] This chapter has sought

---

136   OLCA, "Corte de Apelaciones de Copiapó paraliza proyecto minero El Morro," November 22, 2013, olca.cl/articulo/nota.php?id=103800.

137   Mies, *Patriarchy and Accumulation.*

138   Ibid.

to argue that the aspiration that drove Mies's work can illuminate the implications of the international divisions of labor in late-capitalist development. Despite being materially sutured together by extraction, the subaltern groups of resource-rich nations and those of manufacturing centers are mostly unaware that, despite geographical and cultural difference, their plight is essentially the same. Both face modalities of social proletarianization that actively harness racism, sexism, and nationalism, to fragment labor and drive down wages. Millions upon millions of peasants—on both sides of the Pacific Ocean—have been torn from their families and villages in recent decades in order to be integrated into a globalizing circuit of temporary, racialized, and migrant labor.

If there is no insistence on these relations of coexistence and material intercourse, radical thought and action will only tend to tackle the culturally specific manifestations of social domination, not its underlying foundation in the production and reproduction of social life. By describing the new laboring subjects that populate the worlds of East Asian manufacturing, especially the Chinese working class, I have sought to demonstrate that late capitalism tends to homogenize the conditions of workers in primary, secondary, and tertiary sectors across the international division of labor. Because the social composition of the global working class has shifted toward a more "plebeian," racialized, and "motley" basis, this chapter has also argued for the urgency of excavating the emancipatory potential of indigenous, *campesino*, and migrant politics. As Ferguson and McNally rightly assert, struggles framed around race, gender, and policing are as strategic as those about workplace issues and constitute a fetishized expression of class struggle in its broadest sense.[139] I have thus argued that because Latin American societies have always grappled with "informality" and the nonsequential coexistence of formal and real subsumption of labor to capital, Marxist scholarship in the region has much to offer to contemporary theorizations of the sociocultural composition of the global working class, beyond the old models and blueprints of Eurocentrism.

For several Latin American intellectuals in the Marxist tradition, asserting indigenous, gender, *campesino*, or other subaltern identities is by no means antithetical to the collective constitution of an overarching materialist, class-based project—a vision that is eloquently encapsulated

---

139   Ferguson and McNally, "Precarious Migrants."

in the ideas of *ch'ixi* modernity and the universal *ayllu*, discussed in the preceding pages. The endlessly destructive—but at the same time endlessly generative—powers unleashed by the technological systems of primary-commodity production have created the material conditions for the *ayllu* to shatter its territorial boundaries and project itself as the springboard for a future form of social organization. By pointing out how some Latin American authors have viewed class struggle through the prism of ethnicity, *mestizaje*, and indigeneity, this chapter contributes to the lively scholarly discussion on the late Marx's interest on premodern and non-Western societies.

Although the forms of social resistance emerging at both ends of the mining supply chain increasingly assume a global and anticapitalist character, ideologies of difference and cultural essentialisms have undermined possibilities for more definitive scale-jumping maneuvers. There have been subtle changes, however, and although some peasant communities are starting to transcend their immediate cultural context to articulate their grievances in idioms of class, such a shift is still in its infancy. For this reason I have attempted to develop a materialist theory of revolutionary subjectivity that deciphers the origin of subaltern political agency in the immanent motion of capitalist technologies of extraction, not in cultural attributes, ahistorical moral imperatives, or a residue of abstractly free human essence that has somehow eluded the grasp of commodification. As Neil Smith long ago argued, human beings have already industrialized all the "nature" that has become accessible to them (including the various configurations of human consciousness) and so to wish otherwise amounts to mere nostalgia.[140] By drawing from debates within the traditions of critical thought that inform this book, I argue that an approach that posits the origins of transformative political action as born of the powers of socialized labor can grasp the imprint of universal intercourse in the production of emancipatory consciousness. This could eventually point the way forward to a different universalism, one based not on abstraction but on concrete specificity.

---

140   Neil Smith, *Uneven Development*.

8. EPILOGUE

Toward an Emancipatory Science in the City of Extraction

This is one of the challenges of our time: We lack bridges between those who resist mining in remote places, those who stand up to glyphosate and the agribusiness, and those of us who live in ever more expensive, fenced, and repressive cities. It is a single struggle . . . but the bonds between people in the city and people in the country are not given; it is up to us to build them.[1]

With a technology dependent on the knowledge of the workers the capitalist mode of production would be an impossibility.[2]

Introduction

Just as the locomotive, the automobile, and the computer deeply shaped the economies and societies of previous industrial eras, the smartphone is actively and systematically reweaving the fabric of modern life in the twenty-first century. Although the pristine appearance of this sophisticated hand machine obfuscates the geological evidence of its creation, a

---

1 Enrique Viale, "El Extractivismo Urbano," in Ana María Vásquez Duplat (ed.), *Extractivismo Urbano: Debate para una Construcción Colectiva de las Ciudades* (Buenos Aires: Editorial El Colectivo, 2017).

2 Sohn-Rethel, *Intellectual and Manual Labor*, 123.

closer look at its material composition indicates the urgency of rethinking extraction beyond the mere spatiality of the mine and of the political territory of its national economy—an argument that is at the heart of this book. A late-generation iPhone can contain up to thirty minerals sourced from a multitude of sites across the globe. Depending on the specific model, the list can encompass rare earth metals such as cerium and neodymium, precious metals such as gold and silver, so-called "conflict minerals" such as cobalt and tantalum, and the more "banal" metals and alloys such as copper, iron, and aluminum, among many others.[3] The "one device" that has come to mediate even the most mundane activities of everyday urban life, then, puts us in a relation of direct material intercourse with the networks of territorial infrastructures of the mining supply chain.

But there is much more to the idea of the planetary mine than mere material connection with the distant geographies and individuals whose lives and times are occluded by the fetishism of the commodity. The same global supply chain that puts the smartphone in our pockets champions an unrelenting race to the bottom in labor standards that increasingly dissolves the distinctions between the workers of the city and those of the noncity, as well as those of the global North and the global South. Thus, the smartphone is not only a *product of extraction*. In the inverted world of capital, it often also functions as an *instrument of extraction* in its own right. In fact, it is becoming increasingly evident that smartphones have opened a new frontier of power and commodification by extending the discipline of the workplace in space and time, allowing constant access to physical and digital selves. In a recently published and influential book on new trends in the digital economy, Shoshana Zuboff claims that the most intimate aspects of human experience have in fact become "the objects of a technologically-advanced and increasingly inescapable raw-material extraction operation." Technology companies, she argues, are systematically transforming human activity into "behavioral data streams" that can be mined, packaged, and sold in intricate and highly profitable financial markets.[4]

---

3   Merchant, *One Device*, chapter 2; Jussi Parikka, *A Geology of Media* (Minneapolis: University of Minnesota Press, 2015).

4   Shoshana Zuboff, *The Age of Surveillance Capitalism: The Fight for a Human Future at a New Frontier of Power* (New York: Public Affairs), 17.

For the above reasons, some authors have recently called for "an expanded conception of extractivism," premised on the idea that some of the dynamics and logics of primary-commodity production are rapidly extending to other domains of socioeconomic activity such as finance, real estate, logistics, and the platform economy.[5] Extractive processes, they argue, provide important analytical insights for elucidating the role of rent, primitive accumulation, and extraeconomic force under contemporary capitalism, especially since the Great Recession of 2008. As a 2015 *Guardian* article puts it, fracking has become the "key inspiration" for today's entrepreneurs and everyday urban life has come to resemble more and more the destruction and grime of an oil extraction site. According to the author, city dwellers are being "redefined as frackable units" by hordes of rentiers and capitalists unleashing all sorts of fees, interests, temporary working contracts, and surcharges to "power-hose the last drops of value out of us all."[6]

In his 1970 book *The Urban Revolution*, Henri Lefebvre famously described the explosion of the city as the unrelenting projection of urban fragments—embodied in built environments, infrastructures, institutions, and everyday practices—across the world's rapidly mutating countrysides.[7] Although deeply generative, his theorization over looked the extent to which the logics and dynamics of resource extraction have also informed the process of capitalist urbanization in definitive ways to such an extent that one could refer to the encroaching intensification of another world-historical, totalizing movement: *the explosion of the mine.*

Historians of technology have long been puzzled by the enduring significance of the artificiality and the mechanized milieu of the mine in the development of modern forms of urbanization. As a result, important sociological and historiographical accounts have been devoted to how technologies for lifting, pumping, ventilation, communication, and lighting implemented in deep underground shafts have played central roles in the construction of skyscrapers and urban

---

5   Gago and Mezzadra, "Critique of the Extractive Operations of Capital"; Mezzadra and Neilson, "On the Multiple Frontiers of Extraction."

6   Ian Martin, "Fracketeering: How Capitalism Is Power-Hosing the Last Drops of Value Out of Us All," *Guardian*, June 30, 2015, theguardian.com.

7   Lefebvre, *Urban Revolution*.

infrastructure after being imported into the city.[8] The flat woven-wire
rope that replaced the hemp rope used for hoisting in the Comstock
mines of the US, Brechin explains, was the basis for the iron flywheels
that to this day propel the iconic cable cars of San Francisco.[9]
Technologies for building massive vertical mining structures deeper
into the ground, Graham shows, foreshadowed the urban elevator—
one of the most ubiquitous technologies in contemporary cities.[10] To
cite a more recent example, technologies for autonomous trucks and
lorries already at work in the mining industry are to some extent
informing the science of the self-driving cars currently being tested by
Google, Tesla, and other companies.[11]

The transfer of physical infrastructure and technological know-how
from the site of extraction toward the city, however, is only one of the
implications of the explosion of the mine. Its institutional and governance dimensions are of particular concern. In recent years, real estate
and construction companies have aroused fierce controversy for replicating the frontier culture that has long been associated with the practices of mining and oil companies in remote rural areas. Activists and
academics in Latin America coined the term *urban extractivism* as a
means to designate the evolving forms of displacement, ecological
destruction, and enclosure that have resulted from increasingly aggressive techniques of real estate speculation and urban rent exaction in
densely populated agglomerations.[12]

In proposing to conceptualize the extractive industries in terms of
interconnected systems of infrastructures, this book points toward the
urgency of questioning and making visible such evolving frameworks of

8   Mumford, *Technics and Civilization*; Brechin, *Imperial San Francisco*; Williams,
Notes on the Underground; Bridge, "Hole World"; David Mindell, *Our Robots, Ourselves:
Robotics and the Myths of Autonomy* (New York: Viking, 2016); Graham, *Vertical*.

9   Brechin, *Imperial San Francisco*, 65.

10   Graham, *Vertical*, 369–70.

11   David Crouch, "Descent of the Machines: Volvo's Robot Mining Trucks Get
Rolling," *Guardian*, May 29, 2016, theguardian.com.

12   Rodrigo Hidalgo, Pablo Camus, Alex Paulsen, Jorge Olea, and Voltaire Alvarado,
"Extractivismo inmobiliairio, expoliación de los bienes comunes y esquilmación del
medio natural. El borde costero en la macrozona central de Chile en las postrimerías del
neoliberalismo," *Innsbrucker Geographische Studien, Band* 40 (2016): 251–70; Ana María
Vásquez Duplat, "Extractivismo urbano y feminismo: dos claves para el estudio de las
ciudades," In Vásquez Duplat (ed.), *Extractivismo Urbano*; Viale, "El extractivismo
urbano"; Andreucci et al., "Value Grabbing."

material interdependence. Taking the cue from Labban's call to "deterritorialize extraction," the preceding chapters have provided empirical and conceptual elements to make sense of the sprawling circulatory infrastructures that connect sites of extraction to megacities, factories, financiers, fleets of dry-bulk carriers, technocrats, precarious migrants, and industrial workers.[13] Although mine shafts and pits are fundamental elements in the geography of resource-based industries, I have intended to show that these are merely the starting point of an intricate system of generalized transnational intercourse that brings together ecosystems, workers, and cities, into a complex unity under the spatial division of labor. The deterritorialization of extraction that this book has attempted to unfold, however, has been concerned with the dynamics of raw materials production narrowly considered. In my view, the physical deterritorialization of primary-commodity production could nonetheless point the way to future research that interrogates the imprints of extractive logics in other domains of social existence, such as the service industry, real estate, logistics, finance, and digital technology.

The conceptual deterritorialization of extraction thus remains an open project, one that is beginning to be grappled with in productive ways.[14] However, I agree with Mezzadra and Neilson that the increasing salience of the extractive attributes of capital does not necessarily mean that "extractivism" or "neoextractivism" should be the proper name for the contemporary economic system.[15] Although extractive processes provide important analytical elements for elucidating the role of rent, primitive accumulation, and extraeconomic force—among other categories—in the reproduction of socioecological life, it is risky to define the present geoeconomic framework on the basis of a single overarching logic. Too much emphasis on extractive logics can obfuscate the equally relevant function of labor exploitation, impersonal compulsions,

---

13   Labban, "Deterritorializing Extraction."

14   Verónica Gago, "Financialization of Popular Life and the Extractive Operations of Capital: A Perspective from Argentina," *South Atlantic Quarterly* 114, no. 1 (2015): 11–28; Joshua Kirshner and Marcus Power, "Mining and Extractive Urbanism: Postdevelopment in a Mozambican Boomtown," *Geoforum* 61 (2015): 67–78; Sassen, "Predatory Formations"; Gago and Mezzadra, "Critique of the Extractive Operations of Capital."

15   Mezzadra and Neilson, "On the Multiple Frontiers of Extraction."

fetishization, and all those economic processes that—despite not being immediately associated with extraction—are also central to it. For this reason, the book is anchored to a revitalized critique of political economy that posits value as the pulsing engine that drives the process of environment-making in contemporary society.

Toward an Immanent Critique of Extraction

By placing value at the center of an expanded reading of extraction, this book has been driven by the aspiration to make sense of the social and ecological ramifications of advanced mechanization and vertical integration in resource-based industries. The technological sophistication of the robotized trucks, dry-bulk carriers, port cranes, lorries, drills, and shovels being currently harnessed to boost mineral extraction is, for that reason, both frightening and awe-inspiring. Due to the treadmill dynamic of capital accumulation, however, mining companies already consider these technologies insufficient and are developing a new generation of machines that will render formerly inaccessible ore bodies susceptible to commodification. In January 2017, an interviewee at a major mining company mentioned that Codelco and Rio Tinto are currently engaged in a collaborative project to develop a "megabulldozer" with the capacity to operate in conditions of high altitude, zero visibility, and inclement weather. Altitude has been a definitive limit for mining expansion, since some of the heftiest mineral deposits are still locked away atop the towering peaks of mountain ranges, where atmospheric and environmental conditions are unfit for human laborers and (most) machines. Accordingly, this autonomous megabulldozer is intended to combine state-of-the-art robotics with infrared, remote sensing, and laser technologies to open a whole new frontier of extraction in the summits of the Andean mountains.[16]

Although this robotic megabulldozer is still just a prototype, a new generation of autonomous and remotely operated underwater vessels and machines are already targeting mineral ores on the ocean floor, even at 5,000 meters below the surface—opening the door onto a "deep-sea mining gold rush" that could have devastating effects on maritime

---

16   Interview with a Chilean mining company engineer, January 10, 2017.

ecosystems.[17] Exploration licenses for areas of up to 72,000 square kilometers have already been granted to mining companies operating in the western Pacific Ocean, and a firm named Nautilus Minerals has developed giant seabed-crawling robots to extract rocks rich in copper, zinc, and gold at a rate of over 3,000 tons per day.[18] Some of these underwater robots and seabed crawlers are so advanced that the engineers who operate them say "they are like Transformers."[19] All of this, of course, appears rudimentary and parochial compared to recent initiatives for mineral prospecting and extraction in outer space. Led by NASA and a company named Planetary Resources that brings together tech tycoons of all stripes, this lavishly Promethean endeavor involves using robotic spacecraft of extraction to mine near-Earth asteroids rich in minerals.[20]

That the world-making capacities of these advanced machines and robots are not employed for the conscious, democratic, and collective regulation of social metabolism but are mere modes in which the self-expansion of value becomes actualized warrants serious and renewed engagement with the notion of alienation in Marx's mature critical theory. A focus on alienation allows understanding why collective social praxis becomes recast as political-economic forms that present themselves as thing-like, unchangeable, and beyond human control. By incorporating living labor into lifeless objectivity, Marx argues, "the capitalist simultaneously transforms value, i.e., past labour in its objectified and lifeless form, into capital, value which can perform its own valorization process, an animated monster that begins to "work" "as if its body were by love possessed."[21] The extent to which instruments of production have become determined as modes of existence of these uncanny and destructive forces, however, is blissfully ignored by the proponents of the idea of the fourth industrial revolution. In a bewildering yet unsurprising statement, Elon Musk—the magnate, inventor, and

17   Damian Carrington, "Is Deep Sea Mining Vital for a Greener Future—Even If It Destroys Ecosystems?," *Guardian,* June 4, 2017, theguardian.com.

18   Carrington, "Is Deep Sea Mining Vital?"

19   Ibid.

20   Andy Greenspan, "Precious Metals in Peril: Can Asteroid Mining Save Us?," *Science in the News,* Harvard University, October 25, 2016, sitn.hms.harvard.edu; Alan Boyle, "Why Asteroids Loom as a Future Space Frontier for Mining and Manufacturing," *GeekWire,* July 31, 2017, geekwire.com.

21   Marx, *Capital,* Vol. 1, 302.

⸺ of technology firms SpaceX and Tesla—recently claimed that artificial intelligence needs to be regulated before we see "robots going down the street killing people."[22] In spaces of extraction, as I have illustrated in detail, robots and other technological systems of extraction are already killing, injuring, and displacing people in manifold ways, while also obliterating extrahuman natures.[23]

In making visible the fantastical, fetishized forms spurred by advanced iterations of industrial technology, this book has not been exclusively intended as a counterpoint to the now popular idea of the fourth industrial revolution. Most importantly, I have also envisaged it as a springboard for comradely dialogue with other critical readings of resource extraction and socioenvironmental conflict in Latin America and beyond. The traditions of form-analysis Marxism that underpin this book, coupled with recent readings of the new international division of labor, provide important elements for "reproducing with thought" the evolving processes that produce sites of extraction in historically and geographically specific ways. Attributing the dynamics of primary-commodity production *exclusively* to relations of unequal exchange between core and peripheries, this book has argued, amounts to the sort of political reductionism that is oblivious to the sociomaterial content of seemingly extramundane economic categories. Moreover, such a methodological orientation also results in a structural-functionalist reading of the state as an ahistorical, supratemporal black box, not as the apotheosized manifestation of contingent, sensuous social practice—or "the

---

22 Cited in Olivia Solon, "Killer Robots? Musk and Zuckerberg Escalate Row over Dangers of AI," *Guardian*, July 25, 2017, theguardian.com.

23 In 2017, Elon Musk joined a group of AI experts calling on the United Nations to ban autonomous weapons. Such lethal autonomous weapons, these experts warned, could be "weapons of terror, weapons that despots and terrorists use against innocent populations, and weapons hacked in undesirable ways" (Peter Holley, "Elon Musk Calls for Ban on Killer Robots before 'Weapons of Terror' Are Unleashed," *Washington Post*, August, 21, 2017, washingtonpost.com, accessed August 28, 2017). However, Musk seems to be unaware that the very technologies for self-driving vehicles he is currently pioneering could become massively weaponized by the same truck drivers and taxi drivers left unemployed by them. According to MIT Technology Review, out-of-work truckers armed with "adversarial machine learning" could dazzle autonomous vehicles into crashing at highways and streets, creating widespread chaos and becoming a crucial obstacle to the development of self-driving vehicles. Simson Garfinkel, "Hackers Are the Real Obstacle for Self-Driving Vehicles," *MIT Technology Review*, August 22, 2017, technologyreview.com.

concentrated force of the relations of bourgeois freedom," to use Bonefeld's insightful formulation.[24] The materialist theory of the state that this book has proposed, then, assumes particular intellectual urgency as the ravaging powers of globalization reveal in full the antinomies of the concept of the nation-state as the proverbial "one national boat" of relatively homogeneous interests.

The political context of the new international division of labor, it has been argued, has cast into stark light the fact that the *primus motor* that shapes the space economy of extraction is not the political relations of the state; it is the reproduction of the class relations that lie at the heart of the historical movement of modern society. Although the concrete materiality of this process is eminently global, it nonetheless assumes phenomenal reality in the distorted and "celestial" forms of international politics. Most specifically, it is the expanded reproduction of capital, through constant leaps and bounds in the sociotechnical organization of the labor process, which asserts itself as the foundation for the geographies of primary-commodity production. It is the voracious hunger for the minerals, foodstuffs, and other raw materials demanded by East Asia's gargantuan industrial metabolism—not a political quest for "global dominance" nor regulatory strategies—that has driven the expansion of primary-commodity frontiers in twenty-first-century Latin America. It has therefore been an important contention of this book that the rise of China and other Asian Tigers brings the pressing need to conceptualize capitalism as a sociomaterial system whose extent and complexity puts into question major geopolitical imaginations and conceptual frameworks.

Generally speaking, the shift from the global to the planetary necessarily demands a process of conceptual experimentation that can transcend the limitations posed by the intrinsic Atlanticism, state-centrism, and Eurocentrism of traditional social theory, critical and otherwise. As authors in the traditions of Open Marxism and related approaches remind us, categories of analysis should never be taken for granted. The fixity of structural and teleological-determinist Marxism, they argue, should be replaced by an open, exploratory, ongoing interrogation of the ever-shifting configurations of sensuous practice. For Postone, if immanent critique is to set out from the historicity of things, it cannot

---

24   Bonefeld, *Critical Theory*, 184.

be rooted in a standpoint that is extrinsic to the social universe it proclaims to observe. It must be radically embedded within its own context.[25] The materialist critique is *immanent critique* because "it cannot take a normative position extrinsic to that which it investigates" (which is the context of the critique itself). Indeed, Postone continues, "it must regard the very notion of a decontextualized, Archimedean standpoint as spurious."

It is remarkable that Giovanni Arrighi—one of the foremost architects of the grand tradition of world-systems analysis— took very seriously the implications of a new Asia-centered capitalist world system, especially at a time when most of the work in global political economy and critical social theory remained tethered to Atlanticist parochialisms. It is particularly noteworthy that he was reflexive enough to realize that this world-historical shift did not fit into the inherited conceptual frameworks of the past—some of which he had even created. In *Adam Smith in Beijing*, published shortly before his death in 2009, Arrighi explicitly warns about the uncritical extrapolation of fixed templates to describe the new geoeconomic reality. In Arrighi's words, "attempts to foresee China's future behavior vis-à-vis the United States, its neighbors, and the world at large on the basis of the past experience of the Western system of states are fundamentally flawed." For one thing, "the global expansion of the Western system has transformed its mode of operation, making much of its past experience irrelevant to an understanding of present transformations."[26] By positing the question of the organic composition of capital as the immanent content that underpins the international political dynamics of resource extraction, I propose a reading of global political economy that is sensitive to the multiple interdependencies that constitute the geographies of the extractive industries under late capitalism.

Making analytical distinctions between the essential content of capital (i.e., the movement of value) and its historical, sensuous, and phenomenal manifestations (i.e., profit, rents, politics), makes clear that it is not the political relations between states that need to be transformed but the wage system that acts as the foundation for such relations. This insight is relevant, especially when old theories of dependency,

---

25   Postone, *Time, Labor, and Social Domination*, 87.
26   Arrighi, *Adam Smith in Beijing*, 9.

regulation, and unequal exchange are being eagerly brought back to explain China's rising influence in Latin America.[27] In bypassing the domestic context of labor exploitation, Pradella notes, state-centric approaches of that sort tend to consider the state as ultimately benign.[28] The end result, she argues, is that successful cases of industrial development—such as South Korea and China—are construed as models for "all developing countries" to follow. This has led to the view that Latin American nations will manage to overcome their problems when they shed their status as exporters of raw materials and enter the select clique of "cores" or "semiperipheries."

Such intellectual closure, which accepts the horizons of modern society as its own theoretical horizons, encapsulates what Bonefeld and coauthors refer to as a *crisis of structures*, and especially of the crisis of approaches that take structures as their methodological premise.[29] Determinist theory, they point out, becomes complicit in foreclosing the possibilities of a contradictory world. The starting point of a critical theory of extraction should not be the existing conceptuality of society (manifested in ossified categories of analysis inherited from previous industrial eras) but the contingency of actually existing social relations. The capitalist form of wealth is underpinned by the permanent reproduction of a class of dispossessed producers of surplus value and thus, I have argued, needs to be placed squarely at the center of scholarly efforts at understanding resource extraction and uneven geographical development broadly considered. As the boundary that divides manufacturing and extraction becomes increasingly tenuous, radical thought and action against *extractivism* should be redefined on the basis of *total struggle* against capital. The fragmentation of class relations into distinct but interlocking political forms brings with it the atomization of the laboring classes. Individuals are categorized as tenants, peasants, citizens, debtors, or students, and what should be a single struggle against class domination is broken into a cacophony of multiple voices speaking

---

27   Ariel Slipak, "América Latina y China: Cooperación Sur-Sur o 'Consenso de Beijing'?," *Nueva Sociedad* 250 (2014): 102–13; Bolinaga and Slipak, "El Consenso de Beijing y"; Schmalz, "El ascenso de China "

28   Pradella, "New Developmentalism." For a critique of recent reappropriations of Latin American structuralism, see Grigera, "Conspicuous Silences."

29   Bonefeld et al. (eds.), *Open Marxism*.

their own vernaculars.[30] For this reason, the struggle to build up class organization should be a struggle *within* and simultaneously *against* these fragmented forms. As chapter 7 explained, territorial struggles against extraction have already pushed forward the frontier of political possibility in important ways. It is time to escalate these emergent forms of popular power into new political and sociotechnical configurations.

Spaces of Extraction and the Prospects for a Postcapitalist Science

The footage of Máxima Acuña receiving the 2016 Goldman Environmental award,[31] clenched fist raised as she narrates, in electrifying verse, her struggle against large-scale mining in Peru, has come to symbolize an increasingly recurrent phenomenon. Humble villagers who can barely read and write are all of a sudden imbued by forces that propel them to take giant transnational corporations to court, file complex legal mechanisms before multilateral organizations, and jeopardize investment projects worth billions of dollars.[32] At first sight, the

---

30   Holloway, "State and Everyday Struggle"; Ferguson and McNally, "Precarious Migrants."

31   Acuña was awarded the Goldman Environmental Prize (known as the "Green Nobel") in 2016 for successfully stopping the development of Conga, a large-scale copper and gold mine in the Cajamarca region of Peru. Instead of giving a speech, Acuña sang to the audience about how the engineers and militarized security contractors of Newmont and Buenaventura—the companies granted concessions to develop Conga—set her hut on fire and stole her sheep after trying to kill her twice. Acuña took major mining companies to the Peruvian courts, where her lawsuits met favorable rulings by the judiciary. She also spearheaded an international campaign against mining in Cajamarca, which led the Inter-American Commission on Human Rights and the Organization of American States to formally request the Peruvian government to adopt precautionary measures for forty-six leaders of *campesino* and indigenous groups in 2014.

32   Another paradigmatic example is perhaps the case of Luis Yanza, the indigenous leader who led the multibillion-dollar class action lawsuit against oil giant Chevron-Texaco for massive ecological plunder following oil spills in the Ecuadorian Amazon, working with a team of international lawyers and activists and with a coalition of villages and indigenous peoples. Yanza's initiative resulted in a court ruling that ordered Chevron to pay $18 billion in damages to 30,000 indigenous plaintiffs. In 2008, Yanza received the Goldman Environmental Prize ("Pablo Fajardo Mendoza and Luis Yanza," goldmanprize. org/recipient/pablo-fajardo-mendoza-luis-yanza, accessed August 22, 2017.

multiplication of the vital capacities of these individuals would seem to spring from transhistorical elements of social life, such as particular cultural or ethnoracial attributes, or even from transcendental moral imperatives. However, a closer look reveals that the oppositional powers of these men and women are born from the framework of material interdependence begotten by commodity-determined practice. I have shown that the expansion of primary-commodity frontiers across the erstwhile countrysides of Latin America transforms formerly free peasant populations into personifications or "character masks" of economic categories (either as semiproletarian workers, wage-laborers, debtors, or as members of the relative surplus population). However, it is in the context of their alienated existence as bearers of capital's powers that they undergo a radical expansion in their own material powers.

The Zapatista uprising of 1994 in Mexico is perhaps the archetypal example of how modern science (objectified in computers, electronic devices, and digital platforms) and vernacular science (objectified in the material practices of the peasant and indigenous communities) can fuse into a unitary emancipatory logic.[33] In the *Paris Manuscripts*, Marx challenged the ontological distinction between manual and intellectual labor—and thus between human and machine—by claiming that "the idea of one basis for life and another for science is from the very outset a lie."[34] However, as Alfred Sohn-Rethel's landmark intervention demonstrates, modern science has been premised from its very inception on reproducing and enforcing a radical separation between those two domains of social existence.[35] The very stability of capitalism as a social system, Sohn-Rethel argues, came to depend largely on maintaining a sharp divide between the technical and organizational intelligentsia (engineers, scientists, economists, financiers) and the manual workforce, as well as between machines and the people who use them. The automation and semiautomation of machinery concomitant to the third industrial revolution, according to Mandel, led to a further

---

33   Harry Cleaver, "The Zapatistas and the Electronic Fabric of Struggle," in John Holloway and Eloísa Peláez (eds.), *Zapatista! Reinventing Revolution in Mexico* (London: Pluto Press, 1998); John Holloway and Eloísa Peláez, "Introduction," in Holloway and Peláez (eds.), *Zapatista!*

34   Marx, *Economic and Philosophic Manuscripts*.

35   Sohn-Rethel, *Intellectual and Manual Labour*.

entrenchment of such a cleft and hence to mass overspecialization and commodification of intellectual labor-power.[36]

The radical bifurcation between intellectual and manual labor—and thus between science and life—has of course reached its most advanced iteration with the new breeds of intellectual laborers underpinning the current technological revolution. They range from the high-flying engineers who receive salaries comparable to those of "professional athletes" to the legions of underpaid researchers, adjunct professors, and research assistants at universities and nonacademic scientific institutions. Although internally polarized by income and working conditions, the productive attributes of this organ of the collective laborer are the exact antithesis of those of the workers in charge of manual tasks—as chapter 3 illustrated in detail. The foundations of the division between sensuous experience and rational abstraction, however, are brittle and unstable. Some intellectual workers have begun to question the nonsense and futility of contemporary science, especially of a model of scientific knowledge that subordinates the totality of their skills, capacities, and potentialities to the immediate needs of profit-making. As Dussel argues, "insofar as knowledge is not exerted as critical actualization of the consciousness of living labour, dominated class, historical people, it is an elitist science, academic, fetishized, sterile, unnecessary: 'knowledge for nothing.'" As a result, many intellectual laborers have come to realize that only when knowledge becomes "consciousness, class consciousness, people's consciousness"[37] is it true knowledge, science as history.

Although the hacker, the whistleblower, the insurgent architect, and the rebellious student are the most visible figures of the ongoing revolt against neoliberal science, they are just the tip of the iceberg. In geographies of extraction, rogue scientists and engineers have forged alliances with *campesino* and indigenous peoples in their ongoing struggles against large-scale resource extraction. Geologists and geophysicists, for example, provide training and physical equipment for indigenous communities to measure air and water pollution in their territories and thus support their lawsuits and other legal mechanisms with

---

36 Mandel, *Late Capitalism.*
37 Paysa, "Salary Rank," 204.

scientifically reliable evidence.[38] As a result of these ongoing dialogues and cross-fertilizations, indigenous peoples have also become proficient in using drones to monitor and expose the destruction of forests, water sources, and ecosystems by oil and mining companies.[39] Coalitions between artists and local communities have given momentum to struggles against extraction.[40] Struggles against transgenic soybean production have led to their own rubric of insurgent science, one that grows rapidly under the banner of the Red de Científicos Comprometidos (Network of Committed Scientists), which incorporates scientists in Argentina, Mexico, Ecuador, Costa Rica, and Brazil.[41] This "pueblo-centric" model of scientific investigation, Feeney-McCandless notes, condemns the pervasive effects of glyphosate and the political-economic system of pillaging and dispossession in which it is imbricated, as well as, crucially, the forms of knowledge that are complacent with it.[42]

The aforementioned synergies between vernacular knowledge and scientific knowledge are not framed within the patronizing, deafening monologue of modern science. Neither do they result exclusively in an oppositional stance toward the world. The so-called agroecological revolution, which has emerged as an alternative to the dominant agroindustrial model of global food production, is perhaps the best example of how hand, heart, and mind engage in truly dialectical co-determination and invent new modes of regulating the process of social metabolism. According to Altieri and Toledo, agroecological methods of food production do not amount to the sort of pastoralism that naively romanticizes premodern peasant societies.[43] On the contrary, they constitute a scientific-technological approach to food production that combines the

38   Interview with Julio Fierro, director of Terrae, September 26, 2013; interview with Mark Muller, geophysicist from the London Mining Network, September 11, 2015.

39   Irendra Radjawali, Oliver Pye, and Flitner Flitner, "Recognition through Reconnaissance? Using Drones for Counter-Mapping in Indonesia," *Journal of Peasant Studies* 44, no. 4 (2017): 817–33; Reuters, "Panama's Indigenous Tribes Launch Drones to Fight Deforestation," June 2, 2016, reuters.com; *Marabunta*, "Indígenas aprendieron a usar drones para luchar contra la deforestación," June 2, 2016, marabuntaonline.com.

40   Maristella Svampa, "Art and Politics: Identity and the Art of Unbecoming a Colony," *Harvard Review of Latin America* (Fall 2011): 84–85; Marcela Pulgar, "Artistas se reunieron en Alto del Carmen para proteger el agua y los glaciares," *Futuro Renovable*, May 21, 2013, futurorenovable.cl.

41   Feeney-McCandless, "Por una vida digna."

42   Ibid.

43   Altieri and Toledo, "Agroecological Revolution."

millenary and situated knowledges of peasants with the powers of modern science and technology, offering a robust sociotechnical basis for an agrarian revolution that is genuinely sustainable, socially just, and able to feed the planet in times of climate change.[44] This "dialogue of knowledges" (*diálogo de saberes*) forcefully captures the sorts of design interventions that, according to Arturo Escobar, will shape the collective construction of the pluriversal society that will exist against and beyond that of Western, capitalist modernity.[45]

As an idiom of critique, one variety of the pastoral is premised on idealizing rural life or a lost past—real or imagined. But there is another, more oppositional version of the pastoral that interrogates the lives of those in the midst of the maelstrom of modernization to find an implicit future, one that speaks to the anxieties and tensions of the urban present. As chapter 7 explored in detail, Marx spent the final decade of his life on a frantic quest to find such pastoral visions by studying the communistic societal forms of non-Western and premodern cultures. Perhaps his youthful, Feuerbachian reflections on alienation already contained the germ of a pastoral design in which the lost unity between hand and mind could finally be restored, albeit under an advanced technological configuration. In the *Paris Manuscripts*, he claimed that "natural science will in time subsume under itself the science of man, just as the science of man will subsume under itself natural science: *there will be one science.*"[46] The abolition of capital, however, does not necessarily imply that the rift between algorithmic and folk knowledge will be overcome. If such antithesis persists, philosophers of technology warn us, it will ensure the continuation of an antagonistic society—the telos of which is nothing less than nuclear annihilation.[47] For Starosta, the Marxian idea of emancipation is therefore not exclusively premised on the desubordination of natural science to the requirements of the alienated movement of private

---

44 Altieri and Toledo, "Agroecological Revolution"; McMichael, "Peasant Prospects in the Neoliberal Age"; Patel, "International Agrarian Restructuring"; Patel, *Stuffed and Starved*, chapter 7; Peter Rosset and Miguel Altieri, *Agroecology: Science and Politics* (Manitoba: Fernwood, 2017).

45 Escobar, *Designs for the Pluriverse*.

46 Marx, *Economic and Philosophic Manuscripts*, 111. Emphasis added.

47 Mumford, *Technics and Civilization*; Sohn-Rethel, *Intellectual and Manual Labor*; Rose, "Hand, Brain, and Heart."

property; it also fundamentally entails the transformation of the very nature of scientific consciousness.[48]

In fact, the persistence of the antithesis between science and life explains why it is that major socialist experiments throughout the twentieth and twenty-first centuries have been doomed to failure from the very outset. Although a necessary first step, the radical democratization of the forces of production is not enough. As McNally puts it, dialectical reversal not only means the political victory of the oppressed.[49] It also means dereification: restoring sensuous objectivity via the liberation of things, and people, from circuits of abstraction. The fact that our own collective species-capacities have been aggressively torn from us to be inverted into the attributes of machineries of death, impoverishment, and destruction is horrifying. Yet, as Walter Benjamin reminds us, every epoch "not only dreams the one to follow but, in dreaming, precipitates its awakening. It bears its end within itself and unfolds it."[50] The monstrous robots of extraction that today gobble forests, arable lands, rivers, and ocean floors might be *dream images* of the technological landscapes of tomorrow, where humans and machines are no longer character masks of alien forces but work for the *buen vivir* of the whole society. The task ahead, then, is to supersede such mystical and fetishistic forms, realizing the dream images of the present into the pathway for collective, world-historical awakening. Those who struggle against the planetary mine are already taking steps in that direction. It is time for us to meet them, and for them to meet us.

---

48   Starosta, *Marx's Capital*, 34.
49   McNally, *Monsters of the Market*, 267.
50   Walter Benjamin, *The Arcades Project* (Cambridge, MA: Belknap Press, 2002 [1982]), 13.

Index

Cf denotes figure

A

absolute surplus value, 60n93
abstract risk, 184
accountability, 69, 70, 133
accumulation. *See also* primitive accumulation
of capital, role of liberal state within
    dynamics of, 159
by dispossession, 150n38
domestic spheres of, 43
formation of distinctive national and
    regional spaces of, 86
historical cycles of, 110–11
lever of capital accumulation, 86
modalities of high finance of, 177
systemic cycles of, 41, 42, 43
Acuña, Máxima, 254
Adam Smith in Beijing (Arrighi), 252
Africa, "the new scramble for," 70–1
agroecological revolution, 257–8
AI (artificial intelligence), 3, 11, 36, 48, 49,
    118, 250
Alessandri, Arturo, 154
Alianza del Pacífico (Pacific Alliance),
    125–6
Allende, Salvador, 154, 159
alter-globalism, 108
Althusser, Louis, 122n46
Altieri, Altieri, 257
Alto del Carmen, Chile, 235, 236
Angamos, Chile, 130, 134
Anthropocene, 83n25
Anti-Terrorism Statute (Chile), 135
Antofagasta, Chile, 76, 77, 100, 102, 130–1,
    132, 133, 134, 168
Antofagasta Plc., 117
APEC (Asia-Pacific Economic Cooperation),
    119, 124–5, 133
APEC Transportation Working Group (TWG),
    125
Araghi, Farshad, 98, 102, 189
archaic community, according to Marx, 210,
    217, 218, 219, 240
Argentina
mineral prospecting in, 220
mining agreement with Chile, 168–70
mining code in, 166

peasant displacement in, 99
as user of Chinese credit, 186
Arica, Chile, 168
Arrighi, Giovanni, 15, 24, 41, 42, 252
artificial intelligence (AI), 3, 11, 36, 48, 49,
    118, 250
artisan-operator, 104
Asamblea por el Agua del Guasco Alto, 236
Asia
labor unrest in, 225–33
radical politics in, 210
revolutionary consciousness in, 229
share of global manufacturing by, 46
Asian economies, as international lenders, 185
Asian Tigers
emergence/rise of, 10, 12–13, 24n49, 31, 40,
    46, 55, 251
as logistics empire, 116
Asia-Pacific Economic Cooperation (APEC),
    119, 124–5, 133
Atacama Desert, 16, 76, 106, 110, 129–30, 135,
    137, 197
Atlas of Economic Complexity, 14
Austrian School, 149, 160
authoritarian neoliberalism, 128
automation
archetypal principles of, 82
consequence of, 2
in mining industry, 3–4, 36, 49, 248–9
tendency toward as unfolding alongside
    internal polarization between various
    organs of collective laborer, 91
autonomous weapons, 250n23
ayllu, 28, 217, 218, 219, 220, 221, 234, 242
Aylwin, Patricio, 159, 169

B

Backhaus, Hans-Georg, 145
Ballvé, Teo, 174
"banal neoimperialism," 69
Baraona, Pablo, 157
Barrick Gold, 179, 188, 189–94, 198, 198*f*,
    203–4, 235, 239
Bauer, Carl, 163, 164, 165
"Beijing Consensus," 51
Benjamin, Walter, 259

bessemer steel production, consequences of, 43

BHP, 36

binational mining agreement, between Chile and Argentina, 168–70

biogeophysical costs, radical externalization of, 12, 50

blockades, 2, 109, 110, 133, 134, 222, 237, 238

blockages, 133

Bogor Declaration (1994), 125

Bolinaga, Luciano, 51

Bolivia
  mining code in, 166
  model of accumulation in, 216
  percent of surface of available for mineral forecasting, 221
  social mobilization in, 222–3

Bonefeld, Werner, 52, 58, 72, 136, 144–5, 148, 156, 157, 204, 230, 251, 253

Bosteels, Bruno, 217

Boyd, William, 183

Braunmühl, Claudia von, 26, 123

Brazil
  mineral prospecting in, 220
  mining code in, 166
  as user of Chinese credit, 186

Brechin, Gray, 61, 91–2, 246

Brenner, Neil, 43, 143

Bretton Woods gold standard, 177, 189

BRICS countries
  and African countries, 71
  emergence of, 46

Bridge, Gavin, 111, 167

Brigada SOS Huasco, 236

British, ascent of to trade dominance, 43

Bruff, Ian, 128

Büchi, Hernán, 157

Budds, Jessica, 161–2, 163

Buenos Aires-Based Centro para la Investigación como Crítica Práctica (CICP), 54

Bunker, Stephen, 41–2, 43, 110, 114, 116, 185

C

Cáceres, Carlos, 157

Cademartori, Jan José, 94

Calama, Chile, 100, 196

Caldeira, Teresa, 102

Caligaris, Gastón, 58

Cammack, Paul, 9

campamentos, 21, 76, 100, 106, 131–2

Cancino, Arturo, 167

capital
  according to Marx, 38
  according to Postone, 17–18
  accumulation of as global in content and national in form, 138
  accumulation of on world scale, 26
  as alienated subject of modern existence, 210
  antagonism between capital and labor, 228
  circulation of and logistics turn in extractive industries, 113–19
  conflation of with violence, 151
  empire of according to Meiksins Wood, 26–7, 59n91
  as escalating into global socionatural system, 14
  expanded reproduction of, 251
  fictitious capital, 170, 176
  finance and real subsumption of nature to, 182–9
  formal and real subsumption of labor to according to Marx, 60n93
  geo-economic shift in process of global accumulation of, 12
  international movement of, 6
  opportunism of, 142
  organic composition of, 2, 20, 78, 252
  predatory nature of contemporary formations of, 158
  prioritization of in its money form, 192
  real subsumption of nature to, 182–9, 204
  as sociomaterial form of life, 8
  subsumption of humanity to, 80

Capital (Marx), 52, 82, 113, 149n37, 159, 184, 236

capitalism
  casino capitalism, 7
  as emancipated from traditional heartland and global in geographical extent, 9
  financialization of, 176, 177, 189–92
  global capitalism, 22, 30, 40, 57, 71, 106, 203
  as inherently geographical project whose central mediating mechanism is the modern state, 141
  late capitalism, 11, 20, 26, 177, 180, 208, 241, 252
  periodization of industrial capitalism, 40. See also periodizations
  as sociomaterial system whose extent and complexity puts into question major geopolitical imaginations and conceptual frameworks, 251
  think tanks as structuring element of cultural circuits of, 158
  violence of, 142

capitalist expansion, 81, 151

capitalist imperialism, 5, 15, 26, 37, 59, 59n91

capitalist production, 31, 46, 60, 78, 81, 83, 108, 152, 233

capitalist society, 22, 51, 52, 54n75, 71, 81, 85, 149, 173, 218, 231

Carlos Mariátegui, José, 214–15

Carmody, Pádraig, 24, 67, 70–1

casino capitalism, 7

Castro-Gómez, Santiago, 211

Centeno, Miguel, 156
Chapare, Bolivia, 222
Charnock, Greig, 178
Cheng, Jia Ching, 225
Chevron-Texaco, lawsuit against for ecological plunder, 254n32
Chiang Kai-shek Square (Taipei), protests at, 229
Chibber, Vivek, 27, 39, 86
Chicago School/Chicago Boys, 148, 149, 152, 155, 157, 158, 159, 160–1, 168, 172, 191–2
Chile
  Antofagasta as largest port complex in, 130. *See also* Antofagasta, Chile
  attacks on infrastructure in, 110
  constitution of (1980), 152
  copper-mining industry in, 16, 89
  copper production by, 64, 75–6, 166
  creation of new gamut of property owners and rentiers in, 141
  "dirty" forms of energy in, 165
  distribution of services provided to mining industry in, 90f
  dry-bulk cargo in, 117
  emphasis on operational integration in extractive industries of, 119
  energy matrix in, 165
  extraction of copper in, 16
  feminization of migrant population in mining towns of, 101–2
  General Law of Electricity Services (LGSE), 164, 165
  governance of natural resources in, 160–71
  increase in transfer speed at ports in, 119
  institutional redesign and state restructuring in, 127
  labor insurgency in, 224
  Law 18.097, 167
  Law 18.314, 135
  Law 18.985, 169
  Law 19.137, 169
  Law 19.542, 127, 131
  Law 20.773, 131
  maritime cargo in, 127
  mineral exports of, 11, 14
  mineral prospecting in, 220
  mining activity as crucial transformative force in material economy of, 165–6
  mining agreement with Argentina, 168–70
  mining and agriculture as flagship industries in primary-commodity sector of, 161
  Mining Code, 166
  Ministry of Transport and Telecommunications (MTT), 127
  National Law of Mining Concessions (LOCCM), 166, 167–8
  neoclassical economic thought in, 143, 148, 152
  neoliberal technocracy of, 142
  police repression in spaces of extraction in, 134
  port industry in, 129, 132
  privatizations in, 162
  proliferation of retail chains in, 196–8
  ranking in 2015 Fraser Institute's best countries for mining investment, 238
  reduction of port fees in, 128
  rise of neoliberal technocratic rule in, 153–60
  role of state as mediating agent in expansion of physical production in, 191
  seven modernizations of, 160
  technical expertise as fundamental component of policymaking in, 154
  technical innovations in mining industry in, 104
  toll of neoliberal revolution in, 150
  transfer speed in, 128
  Water Code, 161–4
Chilean Christian Democratic Party, 158
Chilean Economic Development Agency (CORFO), 154
Chilean Internal Revenue Agency (SII), 234
China
  balance of trade with Latin America, 14, 63
  Chile's exports of raw materials to, 64
  as commercial partner for Latin American economies, 66, 67
  consumption of world copper production by, 64
  container ports in, 116
  dramatic reconfiguration of infrastructural matrix of, 226
  emphasis on "indigenous innovation" by, 47
  fishing armada of, 116
  foreign exchange reserves in, 185–6
  global presence of, 66
  "Go Out" policy, 185
  growth of cities in, 61–2
  hukou system in, 21, 207, 225
  import of foodstuffs by, 64f
  import of minerals by, 3f, 62f
  imports of raw materials by, 56
  investment strategies of for primary-commodity production, 66–7
  Labor Contract Law, 227
  labor insurgency/unrest in, 225–6, 233
  labor reforms in, 227
  as largest industrial proletariat, 55–6
  loans and lines of credit from to Latin America governments, 186
  as major importer of raw materials, 61
  manufacturing of metals in, 16
  mass exodus toward cities of, 225
  merchant-marine fleet of, 116
  new forms of labor organization in, 225

number of migrants living in urban areas of,
   207
One Belt, One Road initiative, 138
plights of migrant works in, 21
and resource imperialism, 23–4, 24n49
rise of, 48
"robot revolution" in, 47–8
role of, 51
social protest in, 228
special economic zones (SEZ), 225
steel industry in, 65, 117
use of term "migrant" in, 206–7
users of credit from, 186
ch'ixi modernity, 28, 219, 242
Chua, Charmaine, 94–5
Ciccantell, Paul, 41–2, 43, 110, 114, 116, 185
CICP (Buenos Aires-Based Centro para la
   Investigación como Crítica Práctica), 54
CIEPLAN (Corporation of Economic
   Research for Latin America), 158–9
cities
   global cities, 80, 92, 106
   as "inverted mines," 61
   logistical cities, 131
   mines as tending to be geographically
      remote from, 91
   Simmel on nature of, 202
class
   decolonizing of, 211–17
   global working class. See global working
      class
   international working class. See
      international working class
   Marxian notion of, 212
class consciousness, 223, 227, 256
class fragmentation, 20–1
CNC (computer numerical control), 16
Cochabamba, Bolivia, 222
Codelco, 117–18, 169, 248
Collahusai mine, 224
collective agency, 16
collective consciousness, 16
collective laborer, 31, 32, 79, 80, 81–7, 91, 94,
   95, 96, 97, 106, 107, 235, 256
collective power, 82, 204, 213
Colombia
   land-grabbing in, 173–4
   madresolterismo in, 101
   mineral prospecting in, 220
   mining code in, 166
   peasant displacement in, 99, 100
"color codes," for categories of labor-power, 96
Comité Ecológico y Cultural Esperanza de
   Vida, 236
commodity derivatives, 186
commodity fetishism, 121n44
commodity production, 21–2, 81, 208. See also
   primary-commodity production/frontiers
commodity supercycle, 21, 22, 46, 51, 79, 105,

171, 184–5, 195, 199–200, 207, 210, 220,
   237
communes, revival of, 221–2
Communist Manifesto (Marx and Engels), 70
commuting, impact of, 200
computer numerical control (CNC), 16
Comunidad Agrícola Diaguita los
   Huascoaltinos, 236
Comunidad Di aguita los Tambos, 236
Comunidad Diaguita Montañas Fértiles, 236
Comunidad Diaguita Patay Co, 236
Concertación, 159, 168
concession schemes, 171
"conflict minerals," 244
connectivity, according to Khanna, 120
"Connectivity Blueprint," 124–5
connectivity urbanism, 111, 128
connectography, 120
consciousness
   class consciousness, 223, 227, 256
   collective consciousness, 16
   production of as integral part of production
      of general material life, 81
   revolutionary consciousness, 209, 229, 233
The Constitution of Liberty (Hayek), 152
Convention 169 of International Labor
   Organization, 238, 240
cooperative interaction, 82
Copiapó, Chile, 168, 196
CORA (Land Reform Corporation), 154
cores and peripheries, 5, 19, 41, 60, 72, 253
CORFO (Chilean Economic Development
   Agency), 154
corporate financial instruments, impact of, 178
Corporation of Economic Research for Latin
   America (CIEPLAN), 158–9
cosmic megastructure, 57n84
Council for the Defense of the Huasco Valley,
   236
Cowen, Deborah, 25–6, 83, 115, 131
credit
   configuration and reconfiguration of
      geographies of consumer credit, 198
   proliferation of offered by retail stores in
      Chile, 196–7
crisis of structures, 253
Cruzat, Manuel, 157
Cuba, mining code in, 166
Cueva, Agustin, 53, 54
cybertariat, 84
cyborg, 231–2

D
"dark factories," 16
debt
   of extraction, 175–205
   increase of in resource-exporting nations
      and mining companies, 185
   of rural petty producers, 189

as transforming free peasantries into
    indebted proletarians, 188
De Castro, Sergio, 155, 157
Delhi Mumbai Industrial Corridor, 99
De Los Reyes, Julie Ann, 191
De Mattos, Carlos, 92
Democratic Republic of Congo, cobalt and
    tantalum mines in, 68
demonstrations, 227, 237
depeasantization, 79, 97, 99, 102, 104, 105, 107
dependency theory, 37, 43, 44, 44n40, 53–4
dialectical materialism, 211
dialogue of knowledges (*diálogo de saberes*),
    258
difference, universalism of, 217–24
Dussel, Enrique, 44, 52–3, 54
Dutch, ascent of to trade dominance, 42
Dyson, Freeman, 57n84

E
East Asian economics
    industrial expansion and urbanization
        across, 64
    industrial proletariats of, 225
    protest and labor insurgency in, 228–9
    revolution in instruments and relations of
        production in, 46–7
    as world's main creditor nations, 177
Easterling, Keller, 120
Eastman, Agustín Edwards, 157
Echeverría, Bolívar, 102, 212–13, 216
ecological economics, 43
ecological surplus, 50n59
economic abstraction, 172–3
Economic Commission for Latin America and
    the Caribbean (ECLAC), 126, 132–3,
    153–4
economic rationality, 144–53
economic science, 141, 146–7
"the economy," notion of, 147n28
Ecuador
    percent of Amazon basic as concessioned or
        available for concession for oil extraction,
        220–1
    as user of Chinese credit, 186
EGTK (Túpac Katari Guerrilla Army), 215
Ekers, Michael, 107
El Alto, Bolivia, 222
electromagnetic waves, use of, 36
El Morro (gold mine), 240
El Salvador (copper mine), strike at, 1–2, 224
"empire of muddle," 70
ENDESA, 163
Engels, Friedrich, 70, 84
engineered microorganisms, 49
eotechnic phase, 42
Escobar, Arturo, 258
Escondida mine (Chile), 223–4
ethnoracial differentiation, 106

ethnoracial "othering," 87
Europe, downgrading of, 8
existence, modes of. *See* modes of existence
expert idiot/expert idiocy, 33n70, 146
expropriation, technocracy and, 140–74
extended urbanization, 143
extraction
    consequences of technological systems of,
        250
    contemporary organization of, 54
    critique of, 248–54
    debts of, 175–205
    deterritorialization of, 247
    emerging forms of revolt at heart of
        geography of, 229
    indirect forms of dependence as
        fundamental driving force behind spaces
        of, 172
    local communities' encounters with
        technological infrastructures of, 219
    material separation between spaces of and
        spaces of manufacturing, 4, 110
    momentum in struggles against, 257
    in oceans, 248–9
    in outer space, 249
    smartphones as instrument of, 244
    smartphones as product of, 243–4
    spaces of, 87–8
    spatial technologies of, 153
    transformation of formerly free peasants
        into bodies of, 78
    violence of, 173
extractive industries
    logistics turn in, 109–39
    use of water in, 164–5
extrastatecraft, 120

F
factories, "dark factories," 16
faeneros, 94, 100, 105, 106
Falabella, 198
Federici, Silvia, 151
Feeney-McCandless, Ingrid, 257
female workers. *See also* women
    in global South, 208
    labor insurgency in China as being led by,
        226–7
feminism
    standpoint feminism, 7
    territorial feminisms, 223, 226
feminist activities, mobilizing against resource
    extraction, 68
feminization, of migrant population in mining
    towns, 101–2
Feminort (Mining Federation of the North),
    224
Ferguson, Susan, 80, 87, 101–2, 241
fiat moneys/fiat currency, 177, 189
fictitious capital, 170, 176

finance
  and real subsumption of nature to capital,
    182–9
  role of in real subsumption of nature, 188
  urbanization of, 179–82
financial coupon ownership, 187
financial ecology, 197
financial intermediation, 5, 200
financialization
  of culture and experience, 204
  defined, 181
  of everyday life in Huasco Valley, 194–203
  as widely contested notion, 184
Fine, Ben, 146–7
Finning Cat, 90
first technological revolution, 42
Fischer, Karin, 152
flexigemony, 67
floating populations, in mining towns, 94
food consumption, 63
food sovereignty, 28, 108
formal subsumption, of nature, 183, 188
form-analysis Marxism, 6–7, 7n10, 122, 144,
    250
fourth industrial revolution, 12n25, 37, 48–9,
    73, 249
fourth machine age, 4, 10–11, 12, 40, 46–51, 58,
    72, 98
Foxley, Alejandro, 158
fracking, 245
Fraser Institute, 238
free economy, as resting on state's use of
    organized force, 156
Frei, Eduardo, 168
Freiburg School, 149
Freire, Paulo, 179–80
Freirina, Huasco Province, Chile, 236, 239
French, Shaun, 176, 180, 182
Frente de Trabajadores Nelson Quichillao, 3,
    224
Friedman, Eli, 225, 227
Friedman, Milton, 148, 155, 156
Fröbel, Folker, 9
Froud, Julie, 187
Fuentes, Adriana, 166

G
Gallagher, Kevin, 186
García Linera, Álvaro, 104–5, 215–17, 218,
    223, 234
Geist, 17, 18
gender, as element in types of forced
    displacement and migratory trends,
    100–1
General Law of Electricity Services (LGSE),
    164, 165
geological modeling tools, 31, 36, 49, 89
Geopolítica de la Amazonía (García Linera),
    215–16

geopolitical power, transfer of from West to
    non-West, 48
geopolitical relations, new structure of, 37
geospatial information systems (GIS), 4, 11,
    36, 49
Gerber, Julien-Francois, 201–2
Gindin, Sam, 23
Giovanitti, Arturo, 134
global capitalism, 22, 30, 40, 57, 71, 106, 203
global cities, 80, 92, 106
global depeasantization, 79, 97–8, 102
global labor organization, shifts in, 90–1
global manufacturing
  alterations in power dynamics and
    governance composition of, 47
  Asia's share of, 46
global proletariat, construction of, 83
global social factory, 83
global South
  conversion of into "world farm," 50
  female agricultural and industrial workers
    in, 208
  industrialization of, 47n47, 71, 98
  relocation of large-scale industry to, 240
  technological modernization and industrial
    upgrading in, 4–5
global value chain (GVC), 111, 119, 125
global warming, 15, 50
Global Witness, 69
global working class. See also international
    working class
  exploitation of, 57
  growth of, 80–1
  rethinking idea of, 214
  shift in social composition of, 7, 63, 241
  social reproduction of, 87
  sociocultural heterogeneity of, 108
  sociotechnical composition of, 77
Goldcorp, 190, 240
Goldman Environmental Prize, 254n31
gold-mining industry, 190–4
Goonewardena, Kanishka, 180
"Go Out" policy (China), 185
Graeber, David, 28
Graham, Stephen, 61, 246
Greenpeace, 193–4
ground-rent, 170
Grundrisse (Marx), 8, 114
Gutiérrez, Raquel, 222
GVC (global value chain), 111, 119, 125
Gwangju Commune (South Korea), protests
    at, 228–9

H
Hacia un Marx desconocido: un comentario de
    los manuscritos del 61–63 (Dussel), 52
hacienda, 195n90
Hall, Stuart, 73
Haraway, Donna, 145n19, 231

Hardt, Michael, 26, 236
Harvey, David, 54n75, 150n38, 151, 170, 171, 176
Hayek, Friedrich von, 144, 148–9, 152
Hegel, G. W. F., 17, 18, 38, 38n12, 52–3, 211
Heinrichs, Jürgen, 9
Hilferding, Rudolf, 8
History and Class Consciousness (Lukács), 18
Hobbes, Thomas, 175
Hobsbawm, Eric, 97
Holloway, John, 25, 133, 230
Honduras, mining code in, 166
horizontal integration, 128, 223
Huasco, Chile, 110, 135
Huasco Valley, Chile, 177–8, 179, 194–203, 210, 234–40
Huasco Valley Socioenvironmental Movement (Movimiento Socioambiental del Valle del Huasco), 236
hukou system, 21, 207, 225
human biology, reinvention of, 84
human bodies, role of in extraction, 78
human productive subjectivity, 78, 104, 188, 210, 234
human rights abuses, 69
Huneeus, Carlos, 156
Hung, Ho-fung, 48, 68, 228
Huws, Ursula, 84
hydroelectricity, 165

I
Ibáñez, Carlos, 154
idiot savant, 33n70
IIRSA (Initiative for the Integration of South America), 169–70
imperialism
    according to Hall, 73–4
    according to Luxemburg, 60
    capitalist imperialism, 5, 15, 26, 37, 59, 59n91
    as experiential basis underpinning subsumption of constitutive outside of capital to process of accumulation, 73
    as one of phenomenal forms in which global value relations assert themselves, 26
    as political form, 51–61
import substitution industrialization (ISI), 153, 155
income inequality, 94, 105, 131, 137
India, industrialization in, 99
Industrial and Commercial Bank of China, 185
industrial rationality, 105
industrial reserve army, 85, 100. *See also* labor reserve army
industrial revolution
    fourth industrial revolution, 12n25, 37, 48–9, 73, 249
    use of term, 12n25

Iñigo Carrera, Juan, 54, 101, 122n46, 232
Initiative for the Integration of South America (IIRSA), 169–70
Inter-American Court of Human Rights, 238
internal-combustion engine, invention and proliferation of, 43
internal relations, philosophy of, 181
international commerce, defined, 44
International Labor Organization, Convention 169, 238, 240
international working class. *See also* global working class
    fragmentation of productive subjectivity of, 20, 54
    reproduction of, 6
    shift in social composition of, 98
    transformation of modes of existence of, 83–4
interstate system, 19, 24n49, 26, 38, 39, 40, 45, 53, 72
inverted world, 38, 53, 230, 244
Invisible Committee, 121, 134
ion implantation, 50n61
iPhone, as product of extraction, 244
ISI (import substitution industrialization), 153, 155

J
Jameson, Fredric, 11, 182, 217–18
Japan
    ascent of to trade dominance, 43
    international investment by with resource-rich countries and local mining firms, 185
    laboring classes in, 55
    tanker capacities in, 117
Jevons, William Stanley, 147
Junta de Vecinos Chollay, 236
Junta de Vecinos Piedras Juntas, 236

K
Kaltenbrunner, Annina, 186
Kay, Cristóbal, 44
Kerr, Derek, 171
Keynes, John Maynard, 147
Khanna, Parag, 24, 62, 66, 67, 120
Komatsu, 36, 90
Kovalevski, Maxim, 214
Kovalevski Manuscript (Marx), 216
Kreye, Otto, 9

L
Labban, Mazen, 5, 49, 187, 188, 191, 247
labor
    according to Marx, 79
    alienated character of according to Marx, 81–2
    antagonism between capital and labor, 228
    casualization of, 131, 221
    commodification of scientific labor, 84

degradation of, 221
diversity in geographical location of, 91
exploitation of, 80
feminization of, 207n6
flexibilization of, 223
impact of socialized forms of, 209
integration among previously disconnected
    geographies of, 83
interdependence and socialization of, 86
international divisions of, 241
shifting geographies of, 77
Labor Contract Law (China), 227
labor process
    intensification of, 183
    introduction of large-scale systems of
        machinery to, 235
    leaps and bounds in sociotechnical
        organization of, 251
    Marx's treatment of in modern society,
        232n107
    material transformation of capitalist labor
        process, 30, 57, 80n15, 86, 112, 210
    relocation of parts of, 55, 56
    rising capital-intensity in, 233
    robotization and computerization of, 4, 48
labor productivity, 12, 49
labor reserve army, 106. See also industrial
    reserve army
Lagos, Ricardo, 164
land market, 170, 171, 189
Land Reform Corporation (CORA), 154
Lapavitsas, Costas, 196
La pensée marxiste et la ville (Lefebvre), 77
large-scale contractors, shift toward reliance
    on, 89
large-scale industry
    changes in productive technology of, 10
    sociotechnical foundations of according to
        Postone, 27
late capitalism, 11, 20, 26, 177, 180, 208, 241,
    252
Late Capitalism (Mandel), 10
Latin America
    attacks on infrastructure in, 109–10
    balance of trade with China, 14, 63
    changing configurations of resource
        frontiers in, 14, 61
    container traffic in, 117
    debt in, 177, 185, 186
    depeasantization in, 99–100
    emphasis of mining industry in, 111–12
    expansion of iron-ore mines across, 65
    and idea of free economy resting on state's
        use of organized force, 156
    indigenous struggles in, 27
    inflows of cheap migrant labor in, 77
    infrastructure investment in, 220
    interoceanic fantasies in, 170n138
    labor insurgency in, 223

Marxism in, 214
"masculinization of space" in mining towns
    in, 103
messy materialities and entanglements of,
    22
as microcosm of global shift away from
    politics, government, and "dissensus,"
    142–3
mineral exports of, 11
mineral prospecting in, 220
Pink Tide governments in, 29–30, 30n69,
    72
plights of peasants in, 21
police repression in spaces of extraction in,
    134–5
political forces in, 28
port industry in, 119
radical politics in, 210
revolutionary consciousness in, 229
on rise of China, 24
social mobilization in, 221, 223
sociopolitical contestation against resource
    extraction in, 208
as specialized in producing primary
    commodities, 99
Latin American Observatory for
    Environmental Conflicts (OLCA), 221,
    238
Latin American structuralism, 37, 43–4
Law 18.097 (Chile), 167
Law 18.314 (Chile), 135
Law 18.985 (Chile), 169
Law 19.137 (Chile), 169
Law 19.542 (Chile), 127, 131
Law 20.773 (Chile), 131
layoffs, 3, 131, 193, 204
Lefebvre, Henri, 15n36, 77, 130, 177, 178,
    179, 180, 181, 182, 193, 199, 203, 230–1,
    245
Lenin, Vladimir, 8
LGSE (General Law of Electricity Services),
    164, 165
liberal cosmopolitanism, 15
liberalism. See neoliberalism; ordoliberalism
litigation, use of by social movements, 237–8,
    254, 256–7
LOCCM (National Law of Mining
    Concessions), 166, 167–8
Loftus, Alex, 107, 184
logistical cities, 131
logistical connectivity, 112, 129
Logistical Development Program (Chile
    MTT), 127
logistical urbanism, 128–9
logistical urbanization, 32, 111, 112, 128–36,
    137–8
logistics revolution, 31, 114, 115–16, 120
London Mining Network, 221
low-grade mineral deposits, mining of, 49

Lukács, Georg, 18, 235
Luksic Group, 157
Luna, Diego, 169
Lüthje, Boy, 96, 227
Luxemburg, Rosa, 8, 19, 60, 150n38

M
machine age
    fourth machine age, 4, 10–11, 12, 40, 46–51,
        58, 72, 98
    third machine age, 83, 146
    use of term, 12n25
machine rationality, 147, 159
machine regularity, 147
machinery and electrical equipment, total
    world exports of (1988–2015), 57f
machines
    megamachine, 82, 85, 106
    production of motive machines by, 10, 36n8
machinofacturing, 2, 10, 26, 36n8, 50
madresolterismo, 101
Mandel, Ernest, 10, 11, 36–7, 40, 42, 43, 50, 52,
    83, 84, 93f, 146, 255–6
manufacturing
    geographical distribution of increase in
        capacity of, 46
    material separation between spaces of and
        spaces of manufacturing, 4
Manuscripts of 1861–1863 (Marx), 52
Marambio, Alejandro, 197
March, Hug, 184
marches, 237
Marikana mine, strike at, 2n3
Marini, Ruy Mauro, 53, 54
Marx, Karl, 6, 8, 10, 17, 18, 38, 51, 54, 60, 70,
    73, 79, 80, 81, 82, 84, 85, 107, 108, 114,
    121, 122, 123, 149, 150, 151, 164, 172,
    183, 184, 209, 210, 213–14, 216, 217, 218,
    231, 232, 233, 236, 242, 249, 258
Marx-Engels-Gesamtausgabe, 214
material footprint, of robotized mine, 11–12,
    49
material interdependence, 20, 27, 108, 158,
    208, 213, 247, 255
Mbembe, Achille, 8, 78, 102, 228
McMichael, Philip, 108
McNally, David, 59, 80, 87, 101–2, 187, 241,
    259
Medina, Eden, 154
megamachine, 82, 85, 106
megatechnics, 56–7
Meiksins Wood, Ellen, 26–7, 59n91
Mejillones, Chile, 130, 134
Menem, Carlos, 168
Merchant, Brian, 91
Merrifield, Andy, 77
methodological individualism, 160
methodological nationalism, 5, 19, 23, 27, 43,
    44, 50–1, 53, 124, 208

Mexico
    mineral prospecting in, 220
    mining code in, 166
Mezzadra, Sandro, 60, 87, 247
microcredit, 189
microelectromechanical systems (MEMS),
    47n48
microelectronics technological revolution,
    50n61
microfinancial practices, at household level,
    182
Mies, Maria, 21, 78, 208, 240–1
"Miliband-Poulantzas" state debate, 122
Milibrand, Ralph, 122
militarization, 30, 38, 73, 131, 135
Milonakis, Dimitris, 146–7
mine
    cities as "inverted mines," 61
    explosion of, 245, 246
    as interconnected system of spatial
        technologies and infrastructures, 18
    as network of territorial infrastructures and
        spatial technologies vastly dispersed
        across space, 5
    as tending to be geographically remote from
        cities, 91
    as transnational infrastructure, 1
mineral traceability, 118–19
Mining Code (Chile), 166
mining codes, 166–7
Mining Federation of the North (Feminort),
    224
mining industry
    as capital-intensive, 2–3, 12, 49, 64, 175–6,
        221
    consequences of self-objectifying practice of
        capital in, 18
    diversification in services in, 89–90
    horizontal integration of, 223
    impact of on rural lifeworld, 88
    innovations in robotics, biotechnology,
        artificial intelligence, and geospatial
        information systems in, 11
    as mirroring regimes of industrial
        organization in electronics industry, 88–9
    patterns of labor differentiation in, 96
    pressures on to reduce transport costs and
        increase circulation speed of minerals,
        111
    productive articulation between robotics,
        biotechnology, AI, GIS, and control systems
        in, 49
    relentless robotization and computerization
        advanced by, 36
    reorganization of into global supply chains,
        5, 79
    shift toward reliance on large-scale contractors
        in, 89
    as "smart," "flexible," and "autonomous," 3

social mobilization/social resistance against,
  235–6, 237
violence toward peasants and indigenous
  leaders in projects of, 100
mining proletariat, 87–96, 104, `07107
MiningWatch Canada, 221
minor theory, 7
Mirowski, Philip, 146, 147
Mises, Ludwig von, 149
Mitchell, Timothy, 147, 148
modern world system, 44, 45
modes of existence, 6, 8, 17, 18, 52, 63, 71, 83,
  114, 178, 229, 249
money
  as central to metabolism of spatially
    integrated infrastructural systems of
    planetary mine, 175–6
  fiat moneys/fiat currency, 177, 189
  self-generating powers of, 178
monist metaphysics, 38n12
monopoly rent, 170n140
Mont Pelerin Society, 152
Moore, Jason W., 50, 58, 194
Morales, Evo, 215
Movimiento Soci oambiental del Valle del
  Huasco (Huasco Valley
  Socioenvironmental Movement), 236
Mumford, Lewis, 42, 56, 70, 82, 84–5, 88, 106
Munk, Peter, 190, 191, 204
Musk, Elon, 249–50

N
nanotechnologies, 47n48
NASA, 249
national economy, 19, 148, 154, 244
National Law of Mining Concessions
  (LOCCM), 166, 167–8
"National Plan for Port Development" (Chile
  MTT), 127
National Planning Agency (ODEPLAN), 154,
  155, 160
national policy, evolving political forms of and
  international competition, 55
national-scale modes of analysis, 111
nation-state
  according to Braunmühl, 123
  as fetishized form of expression of global
    unfolding of value, 27
  as foremost container of financial activity,
    180, 182
  as internally constituted by its own domestic
    context, 45
  international political relations of, 19, 22,
    26, 30, 71
  political dynamics of, 39
  as proverbial "one national boat," 251
natural-resource curse, 43
natural-resource frontiers
  after Western phase of capitalism, 8–15

expansion and reproduction of, 58, 100,
  216
as internally related to constitution of
  modernity, 40–1
study of, 42, 114
natural-resource governance, 30, 44, 67, 71–2
nature
  externalist conception of, 80n15
  financialization of, 186–7
  formal subsumption of, 183, 188
  production of, 78–9, 81–7, 107
  real subsumption of to capital, 182–9, 204
  subsumption of, 178–9
Nautilus Minerals, 249
Negri, Antonio, 26, 236
Neilson, Brett, 60, 87, 120–1, 247
neoclassical economic thought, 19, 140, 141,
  143, 144, 145–6, 147, 148–9, 152, 153,
  159, 161, 172
neoliberal globalization, 17, 142, 155
neoliberalism, 136n95, 154–5, 156, 159, 165,
  172, 173, 180. See also authoritarian
  neoliberalism
neoliberal technocracy, 32, 142, 144, 149,
  172
Network of Committed Scientists (Red de
  Científicos Comprometidos), 257
Neurda, Pablo, 21
Newcrest, 190
New Dialectics tradition, 149n37
New Gold, 240
new international division of labor (NIDL),
  9–10, 20, 21, 26, 47n47, 54, 73, 208, 250,
  251
Newmont, 190
Nicaragua, mining code in, 166
Nueva Mayoría, 168

O
Observatory for Mining Conflicts in Latin
  America (OCMAL), 221
occupations, 237
ODEPLAN (National Planning Agency), 154,
  155, 160
"Ode to Copper" (Neruda), 21
OLCA (Latin American Observatory for
  Environmental Conflicts), 221, 238
Ollman, Bertell, 181
One Belt, One Road initiative (China), 138
open-cast mining, 69, 76, 239
Open Marxism, 7n10, 178, 230, 251
ordoliberalism, 136n95, 148, 155, 156
Organisation for Economic Co-operation and
  Development (OECD), 126, 132
oriental despotism, 214
Ossandón, José, 155, 196
outsourcing, 92, 94, 188, 204, 208, 223, 227
Owenite socialist thought, 151

P

"Pacific Century," 46

Pacific Ocean, as main infrastructural corridor of world trade, 13, 116

Pajareteros Alto del Carmen, 236

Pakistan, gold, gas, uranium, and oil deposits in, 68

paleotechnic phase, 42

Panitch, Leo, 23

Paraguay, peasant displacement in, 99

Parenti, Christian, 153, 161, 167

Paris Manuscripts (Marx), 81, 255, 258

Pascua Lama, 178, 179, 192–4, 198, 198*f*, 203–4, 235, 239

Pastoral Salvaguarda de la Creación, 236

Patan, Nepal, protests at, 229

patriarchal domination, 73

peaceful rise (*heping juequi*), 67

Pearl River Delta (China), 16, 56, 85, 227

peasantry, twilight of, 97–105

periodizations, 12n25, 24, 40, 42, 43, 50, 72

peripheral urbanization, 102

personal financial instruments, impact of, 178

Peru
    expansion of mineral extraction frontier in, 65
    mineral exports of, 11
    mineral prospecting in, 220
    percent of highland peasant communities affected by mining concessions, 220

picket lines, 237

Piñera, Sebastián, 135

Pink Tide governments, 29–30, 30n69, 72

Pinochet, Augusto, 142, 144, 148, 154, 155, 157, 160, 166

planetary
    Lefebvre on emergence of, 15n36
    shift from global to, 16, 19, 251

planetary mine
    according to Labban, 5
    defined, 4, 16
    foundation of, 140

Planetary Resources, 249

planetary urbanization, 14, 63, 97

pluriverse, 28, 108, 258

police repression, in spaces of extraction, 134–5, 137, 226, 238

political activism, state crackdowns on, 238

political contestation, 109, 221. *See also* sociopolitical contestation

political economy, critique of, 73, 145, 231

political struggle, emerging expressions of, 110

political technology, 143

port operations, standardizing operations and technological systems of, 118

port planning policy, 133

postcolonial studies, 211–12

post-extractivismo, 22, 44

Postone, Moishe, 8, 17, 27, 56, 82, 97, 107, 202, 209–10, 231, 232, 251–2

postwork society, 77n7

Poulantzas, Nicos, 122, 136

power, according to Invisible Committee, 121

Pradella, Lucia, 209, 213, 214, 253

Prebisch, Raúl, 43–4

predatory lending, 197, 201

Prieto, Manuel, 163, 164, 165

primary-commodity frontiers, 22, 25, 32, 239, 251, 255

primary-commodity production, 28, 41, 58, 62, 66–7, 97–105, 107, 110, 114, 119, 177, 204–5, 208, 221, 236, 242, 245, 247, 250

primitive accumulation
    according to Marx, 149, 150, 151, 173
    as constitutive premise of existent economic forces, 157
    new readings of, 149–53, 157–8

production
    forces of, 123n47
    science in immediate process of, 84

productive subjectivity, 20, 54, 78n9, 79, 86, 93, 98, 104–5, 202, 233. *See also* human productive subjectivity

proletarian politics, 223

Proudhonist socialist thought, 151

Prudham, Scott, 183

public health, deteriorations in, 132

Punta Alcalde thermoelectric plant, 239–40

Q

Quichillao, Nelson, death of, 2, 3

R

racialization, 87, 95–6, 101, 208

racialized violence, 25, 73 rational abstraction, 145

raw materials
    access to, 58
    future scarcity of, 186

Razeto, Camilo, 168

real subsumption, of nature, 183, 188

Red de Científicos Comprometidos (Network of Committed Scientists), 257

relative autonomy, 122n46

relative surplus population, 7, 78, 85–6, 172, 255

relative surplus value, 6, 53, 54, 54n75, 56, 58, 60, 66, 233

Ren, Hao, 55–6

rent theory, 153

(re)peasantization, 107–8

research and development, as branch of large companies, 146

resource extraction
    computerization and mechanization of, 234
    contemporary readings of, 142

disintegrating forces as brought about by
     mechanization of, 70
feminist activities mobilizing against, 68
materialist understanding of, 60
relations of social interdependence
     engendered by, 21
sociopolitical contestation against in
     contemporary Latin America, 208
studies of, 124
transnational organization of demands of,
     114
resource frontiers
     expansion of, 61, 65, 68, 100–1
     natural-resource frontiers. *See* natural-
          resource frontiers
resource imperialism
     after the West, 35–74
     and revolutionary subjectivity, 23–8
revolutionary consciousness, 209, 229, 233
revolutionary subjectivity, 27, 32, 209, 210,
     224, 225–33, 242
Riffo Pérez, Luis, 92
Rio Tinto, 248
Rivas, Santana, 100
Rivera Cusicanqui, Silvia, 213, 219
Roberts, William Clare, 151, 152, 159, 173
robotic megabulldozer, development of, 248
Roy, Arundhati, 99
Ruiz, Ruiz, 100
rural indebtedness, 201, 204
rural lifeworld, mining's impact on, 88

S
sabotage, 109–10, 133, 134, 237
Saieh, Álvaro, 157
Saint-Simonian socialism, 151, 157
Salinas, Bárbara, 168, 169, 193
Santiago, Chile
     attractive to highly skilled workers,
          financiers, and executive personnel of
          mining industry, 92
     employment growth in according to
          occupational group, 93f
     percent of mining industry contractors based
          in, 92
Sassen, Saskia, 144, 158, 173
Schmalz, Stefan, 186
Schmid, Christian, 97, 200
Schurman, Rachel, 183
Schwab, Klaus, 48–9, 50, 70
science
     incorporation of in immediate process of
          production, 84
     ongoing revolt against neoliberal science,
          256
     persistence of antithesis between science
          and life, 259
     questioning nonsense and futility of
          contemporary science, 256

as ruled by logic of appropriation, 145
second technological revolution, 43
self-regulating market, 160
semiperiphery, 10, 44, 46, 47, 51, 253
shadow economy, in mining industry, 95
shareholder value, 187, 188, 191, 193, 203, 204
Siemens, 90
SII (Chilean Internal Revenue Agency), 234
Silva, Eduardo, 154
Silva, Patricio, 154, 156, 158, 159
Silver, Beverly, 233
Simmel, Georg, 202
Slipak, Ariel, 51
smartphones
     as instrument of extraction, 244
     as product of extraction, 243–4
Smith, Neil, 35, 78, 80, 81, 84, 107, 178–9, 183,
     188, 242
sociación de Pequeño s Agricultores de San
     Félix, 236
social ills, increase of in mining towns, 94
social mediation, 6, 17, 32, 70, 71, 142, 151,
     222, 231, 237
social mobilization, 136, 221, 222, 235, 237
social movements, 133, 221, 237
social production
     computerization of, 2, 105, 218, 223
     encroaching presence of "informality" and
          labor-intensive regimes of, 214
     material conditions of, 81
     social and organizational basis of, 104
social resistance, 235, 239, 242
social revolt, 221, 222, 226
social totality, 180, 199
Sociedad Química y Minera (SQM), 157
socioecological degradation, 69, 219
socioecological property, state as inseparable
     from process of transforming of, 167
sociomaterial practices, impact of, 202
sociopolitical contestation, 208, 219, 228, 233,
     238–9. *See also* political contestation
sociospatial engineering, 168, 170
Sohn-Rethel, Alfred, 145n18, 255
solid waste, production of by robotized mine,
     11–12
South Africa, "color codes" in, 96
Soviet Diamat, 122, 211
spatial hyperglobalism, 120
standpoint feminism, 7
Starosta, Guido, 54, 57, 82, 83, 84, 86, 232, 258
state
     form-analysis view of, 122
     form of, transoceanic corridors and, 120–8
     as inseparable from process of transforming
          socioecological property, 167
     role of as mediating agent in expansion of
          physical production, 191
state authority, shift toward increasing
     institutional concentration of, 128

state power
  internationalization and concentration of, 124, 138
  and logistics turn in extractive industries, 109–39
  more coercive, centralized, and militaristic configuration of, 112
streamlining, 188, 204
strikes, 1–2, 2n3, 109, 132–3, 134, 226, 230
structures, crisis of, 253
subaltern agency, 212
subaltern fragmentation, 108
subaltern futurism, 213
subaltern groups, 27, 58, 73, 205, 211, 216, 222, 241
subaltern identities, 241
subaltern movements, 223
subaltern political agency, 209, 242
subaltern politics, 210
subaltern social practices, 218
subaltern struggles, 28, 33, 228
subcontracted workforces
  in China, 227
  in mining industry, 89, 94
  plight of, 3
  share of subcontracted workers in Chile's mining industry, 95f
supply chain
  according to Khanna, 120
  China as building global supply-chain empire, 66
  as integrated logistical system, 4
  mapping of, 118
  metabolism of, 13
  militarization of in Chile and Latin America, 135
  mining supply chain, 93, 116, 208, 228, 242
Supply Chain Development Initiative (APEC), 125
supranational zones, 43, 46, 47

T
Taylor, Ian, 24, 67, 171, 191–2
technical disruptions, 133
technical expertise
  as fundamental aspect of governance structures, politics, and policymaking, 142
  as fundamental component of policymaking in Chile, 154
  influence of, 153, 174
technical sabotage, 109–10, 133
Technics and Civilization (Williams), 56
technocracy, and expropriation, 140–74
technological revolutions
  according to Mandel, 36n8, 37
  and cheap energy and cheap food, 58
  first technological revolution, 42
  fourth technological revolution, 12

microelectronics technological revolution, 50n61
  new breeds of intellectual laborers as underpinning current one, 256
  scientific-technological revolutions, 10, 30, 40
  second technological revolution, 43
  third technological revolution, 12, 43, 83
technopolises, 106
temporary contracts/temporary workers, 3, 94, 131, 223, 224, 225
"territorial feminisms," 223, 226
Thammasat University (Bangkok), protests at, 228–9
thermoelectric power generation, 165
third machine age, 83, 146
third technological revolution, 12, 43, 83
3-D visualization methods, use of, 36
Tiananmen Square (Beijing), protests at, 229
tiger-cub economies, 55
Time, Labor, and Social Domination (Postone), 17, 18
Toledo, Victor Manuel, 257
total-cost approach, 115
totality/totalization, 20n41, 177, 178, 179–82, 193, 203
total surplus value of society, 38, 39, 52, 53, 54, 59, 71
trade unions
  Frente de Trabajadores Nelson Quichillao, 3, 224
  harassment of, 135
  social composition of new ones, 224
transatlantic shipping, changes in, 116–17
transcontinental logistical infrastructures, 120
transoceanic corridors, and form of the state, 120–8
transportation
  impact of development of faster, cheaper, more efficient means of, 113
  pressure to reduce costs of, 114
Tsing, Anna, 106
Túpac Katari Guerrilla Army (EGTK), 215
two-century model, 41

U
Ulloa, Astrid, 102, 223, 226
Unidos por el Agua, 236
United States
  changes in payout ratio for oil companies in, 187
  investment strategies of for primary-commodity production, 67
  plights of subcontracted workers in, 21
universal ayllu, 28. See also ayllu
Universidad Católica, 157, 158
University of Chicago. See Chicago School/Chicago Boys
urban environments, making of, 75–108

urban extractivism, 246
urbanization
    according to Schmid, 200
    according to Simmel, 202
    of countryside, 88
    extended urbanization, 143
    fractured and polarized spaces of, 77
    as hinging upon speed at which money
        circulates, 203
    logistical urbanization, 32, 111, 112,
        128–36, 137–8
    peripheral urbanization, 102
    planetary urbanization, 14, 63, 97
The Urban Revolution (Lefebvre), 180, 181, 245
Uruguay, mining code in, 166
utopian archipelago, 218, 219

V
Valemax, 65
Vallenar, Huasco Province, Chile, 194–9, 198f,
    199f, 200, 202, 203, 234, 236
Valparaíso, Chile, 131
value. See also absolute surplus value; global
    value chain (GVC); relative surplus value;
    shareholder value; total surplus value of
    society
    Marxian labor theory of, 113
    movement of, 17
    "treadmill effect" as unleashed by self-
        expansion of, 56
van der Pijl, Kees, 81, 88
Venezuela
    mining code in, 166
    as user of Chinese credit, 186
Vergara Marshall, Ángela, 104
vertical reintegration, 32, 85, 88–9, 119
Viña del Mar, Chile, 152
violence
    of capitalism, 142
    condemnation of yet reliance on, 172
    conflation of capital with, 151
    of extraction, 173
    racialized violence, 25, 73
    in today's mines, 172–3
    toward peasants and indigenous leaders, 100

W
wage labor/wage-laborers, 32, 63, 83, 91, 146,
    149n37, 236, 255
wage system, 11, 78, 102, 105, 252–3
Wallerstein, Immanuel, 19, 41, 43, 44, 47
"Washington Consensus," 51
Water Code, 161–4
water rights, 161, 162, 163, 164–5, 168, 171
wealth
    circling flows of mineral wealth, 77
    crisis in encroaching concentration of, 15
Webber, Jeffrey R., 215
Williams, Rosalind, 56
women. See also female workers; feminism;
    feminist activities; feminization
Asia's increasing reliance on women laborers,
    207
    as leading popular mobilizations against
        extraction, 223
world development, according to Marx, 211
world-ecological revolutions, 50n59
world market
    according to Marx, 51–2, 54
    as analytical starting point from which
        nature of resource imperialism can be
        most adequately fleshed out, 38
    autonomization of, 204
    as sociomaterial system organized in form
        of national economies as its aliquot parts,
        20, 124
    violent dislocations as spurred by, 25
world-systems analysis, 19, 23, 37, 40, 41,
    252

Y
Yangtze River Delta, 227
Yanza, Luis, 254n32

Z
Zapatita uprising (Mexico, 1994), 255
Zasulich, Vera, 214, 217, 218
Zhang, Lu, 233
Zibechi, Raúl, 222
zonas de sacrificio (sacrifice zones), 94, 137
Zuboff, Shoshana, 244